Student Workbook

Wood
TECHNOLOGY & PROCESSES

McGraw Hill Glencoe

Safety Notice

The reader is expressly advised to consider and use all safety precautions described in this guide or that might also be indicated by undertaking the activities described herein. In addition, common sense should be exercised to help avoid all potential hazards.

Publisher and Author assume no responsibility for the activities of the reader or for the subject matter experts who prepared this guide. Publisher and Author make no representation or warranties of any kind, including but not limited to, the warranties of fitness for particular purpose or merchantability, nor for any implied warranties related thereto, or otherwise. Publisher and Author will not be liable for damages of any type, including any consequential, special or exemplary damages resulting, in whole or in part, from reader's use or reliance upon the information, instructions, warnings or other matter contained in this guide.

Internet Disclaimer

The Internet listings in this guide are a source for extended information related to the student text. We have made every-effort to recommend sites that are informative and accurate. However, the sites are not under the control of Glencoe/McGraw-Hill, and, therefore, Glencoe/McGraw-Hill makes no representation concerning the content of these sites. We strongly encourage teachers to preview Internet sites before students use them. Many sites may eventually contain links to other sites that could lead to exposure to inappropriate material. Internet sites are sometimes "under construction" and may not always be available. Sites may also move or have been discontinued completely by the time you or your students attempt to access them.

Credits

All photography by the Kevin May Corporation

McGraw Hill Glencoe

The McGraw-Hill Companies

Copyright © The McGraw-Hill Companies, Inc.
All rights reserved. No part of this publication may be reproduced or distributed in any form or by any means, or stored in a database or retrieval system, without the prior written consent of The McGraw-Hill Companies, Inc., including, but not limited to, network storage or transmission, or broadcast for distance learning.

Send all inquiries to:
Glencoe/McGraw-Hill
4400 Easton Commons
Columbus, OH 43219

ISBN 978-0-07-894095-8
MHID 0-07-894095-8

Printed in the United States of America
10 11 12 13 14 15 16 LHN 19 18 17

Table of Contents

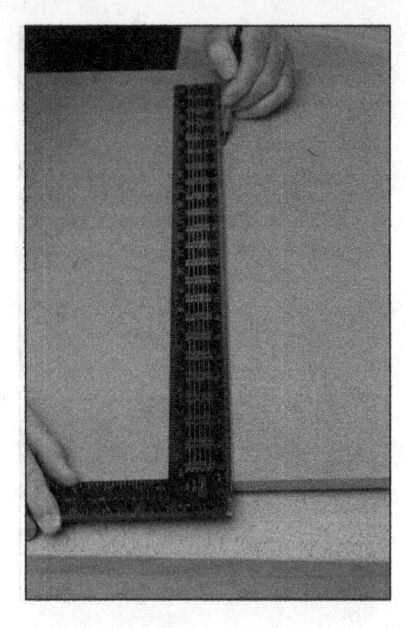

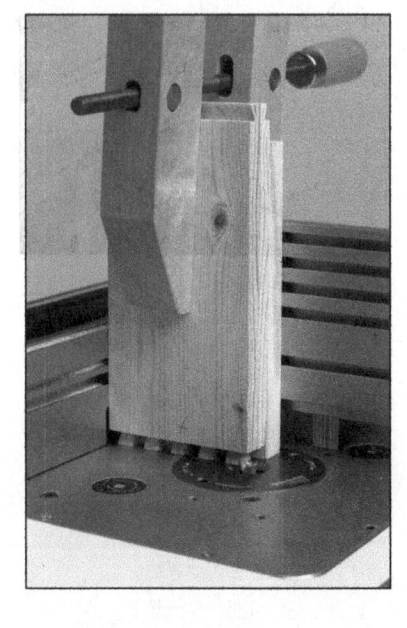

CHAPTER WORKSHEETS 7

Working with Wood
Chapter 1: The Woodworking Industry 7
Chapter 2: Safety Practices 9

Basic Tools and Operations
Chapter 3: Designing and Planning 11
Chapter 4: Measuring and Cutting 13
Chapter 5: Nailing 15
Chapter 6: Drilling 17
Chapter 7: Planing, Chiseling, and Sanding 19

Joinery and Assembly
Chapter 8: Butt, Biscuit, and Dowel Joints 21
Chapter 9: Rabbet Joint 23
Chapter 10: Dado Joint 25
Chapter 11: Lap Joint 27
Chapter 12: Miter Joint 29
Chapter 13: Mortise-and-Tenon Joint 31
Chapter 14: Dovetail Joints and Casework 33
Chapter 15: Using Fasteners 35
Chapter 16: Gluing and Clamping 37
Chapter 17: Installing Hardware 39
Chapter 18: Plastic Laminates 41
Chapter 19: Veneering 43

Using Machines
Chapter 20: Planer 45
Chapter 21: Jointer 47
Chapter 22: Table Saw 49
Chapter 23: Radial-Arm Saw 51
Chapter 24: Band Saw 53
Chapter 25: Sliding Compound Miter Saw 55
Chapter 26: Scroll Saw 57
Chapter 27: Drill Press 59
Chapter 28: Router 61
Chapter 29: Sanders 63
Chapter 30: Lathe 65

Table of Contents (continued)

Finishing
- Chapter 31: Preparing for Finishing........ 67
- Chapter 32: Applying Stains and Clear Finishes........ 69
- Chapter 33: Applying Paints and Enamels........ 71

Construction
- Chapter 34: Preparing for Construction........ 73
- Chapter 35: Sitework and Foundations........ 75
- Chapter 36: Framing and Enclosing the Structure........ 77
- Chapter 37: Completing the Exterior........ 79
- Chapter 38: Completing the Interior........ 81
- Chapter 39: Remodeling and Renovation........ 83
- Chapter 40: Maintenance and Repair........ 85

Advanced Woodworking Techniques........ 87

SCIENCE ACTIVITIES
- Activity 1: Water Evaporation from Wood........ 89
- Activity 2: Testing Wood Preservatives........ 91
- Activity 3: Determining Wood Density........ 93
- Activity 4: Viscosity of Wood Adhesives........ 95
- Activity 5: Muscle Force for Planing........ 97
- Activity 6: Effect of Acid Rain on Plants........ 99
- Activity 7: Analyzing Tree Rings........ 101
- Activity 8: Leaf Structure and Tree Growth........ 103
- Activity 9: Hardness of Sanding Abrasives........ 105
- Activity 10: Identifying Simple Machines........ 107
- Activity 11: How Solvents Work........ 109
- Activity 12: Forces in Wood Framing........ 111

Table of Contents

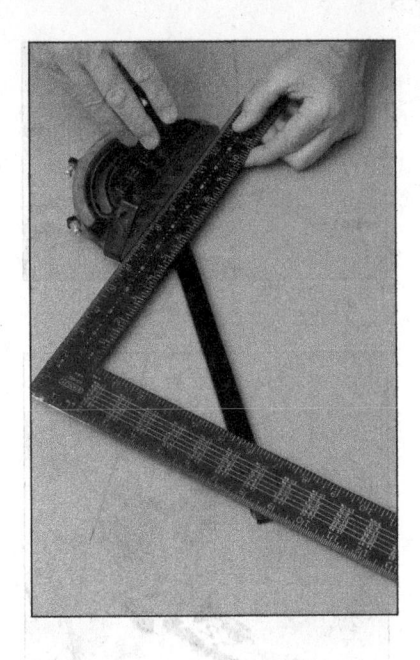

MATHEMATICS ACTIVITIES
Activity 1: Reading a Table 113
Activity 2: Using the Metric System. 115
Activity 3: Reading a Working Drawing. 117
Activity 4: Using a Ruler 118
Activity 5: Fractions and Decimals. 119
Activity 6: Figuring Percentages. 121
Activity 7: Figuring Square and Board Feet 122
Activity 8: Wood Screw Sizes. 123
Activity 9: Finishing Coverage 124
Activity 10: Cost of Electricity for Woodworking Machines 125

SAFETY WORKSHEETS
Workshop Safety 127
Fire Safety 129
Hand/Portable Power Tool Safety. 131
Planer Safety 133
Jointer Safety 135
Table Saw Safety 137
Radial-Arm Saw Safety 139
Band Saw Safety 141
Sliding Compound Miter Saw Safety. 143
Scroll Saw Safety 145
Drill Press Safety 147
Router Safety 149
Sander Safety 151
Lathe Safety 153
Adhesives/Hazardous Materials Safety 155
Handling Finishes Safely 157
Personal Protective Equipment 159

Table of Contents *(continued)*

WOODWORKING ACTIVITIES
Activity 1: Making a Miter Box 161
Activity 2: Building Your Own Sawhorses 162
Activity 3: Using Problem Solving to Help People
　　　　　with Disabilities 164

TIPS AND TECHNIQUES
Tips and Techniques 1:
　　Using Woodworking Tools 165
Tips and Techniques 2:
　　Making Accurate Measurements 168
Tips and Techniques 3:
　　Cutting Dovetails with a Keller Jig 170
Tips and Techniques 4:
　　Making Finger-Lap Joints with an Incra Jig 172
Tips and Techniques 5:
　　Making Half-Blind Dovetail Joints with an Incra Jig 175
Tips and Techniques 6:
　　Making Small Raised-Panel Doors
　　with a Junior Router Set 178
Tips and Techniques 7:
　　Using Clamps 182
Tips and Techniques 8:
　　Making Pocket-Hole Joints 185

WOODWORKING PROJECTS
Project 1: Kitchen Trivet 190
Project 2: Sunny-Day Sandbox 192
Project 3: Housing for the Birds 196
Project 4: Country Chair 200
Project 5: Crown Molding Boxes 204
Project 6: Workbench 210
Project 7: Big Rig for Little Drivers 217

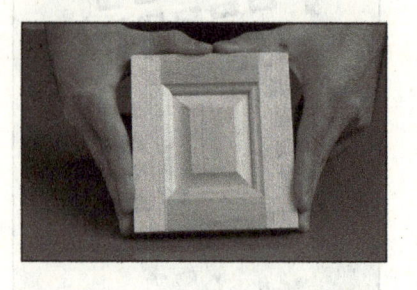

Visit glencoe.com and use
this code to access math activities,
safety quizzes, and other resources.

QuickPass™
WOODTECH2011B1-40

Name _____ Date _____ Class _____

CHAPTER 1: The Woodworking Industry

Matching

Directions: On the line next to each term, write the letter of the correct description.

_____ 1. panel stock
_____ 2. engineered wood
_____ 3. plywood
_____ 4. clear cutting
_____ 5. softwoods
_____ 6. molding
_____ 7. selective cutting
_____ 8. hardwoods

A. narrow strip of wood shaped to a uniform curved profile throughout its length
B. method of harvesting trees in which all trees are removed, regardless of size or type of tree
C. wood that has been processed into sheets
D. wood from deciduous trees such as oaks and maples
E. made from sawdust or small wood pieces and plastics
F. method of harvesting only trees of a certain size and type, leaving all the others behind
G. made by gluing thin layers of wood together
H. wood from coniferous trees such as pine or cedar

Completion

Directions: On the line to the left of each sentence, write the word or phrase that correctly completes the sentence or answers the question.

_____ 9. The first step in processing lumber is ___.

_____ 10. The lumber shown in Fig. 1-1 has been ___-sawed.

_____ 11. Lumber that contains a lot of moisture is called ___ lumber.

_____ 12. Lumber that has been put through a planer is known as surfaced, or ___, lumber.

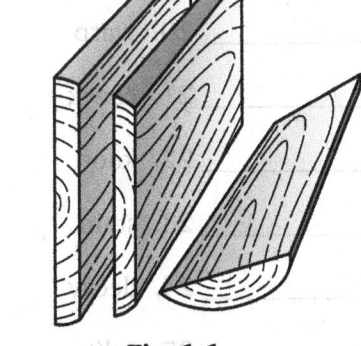

Fig. 1-1

_____ 13. Computer-numerical ___ involves the use of computers and a numerical code to control manufacturing machines.

_____ 14. Computer-aided ___ systems are systems in which a CAD drawing can be sent directly to a computerized machine that makes the product.

(Continued on next page)

7

Name _____ Date _____ Class _____

_____ 15. The lumber shown in Fig. 1-2 has been ___-sawed.

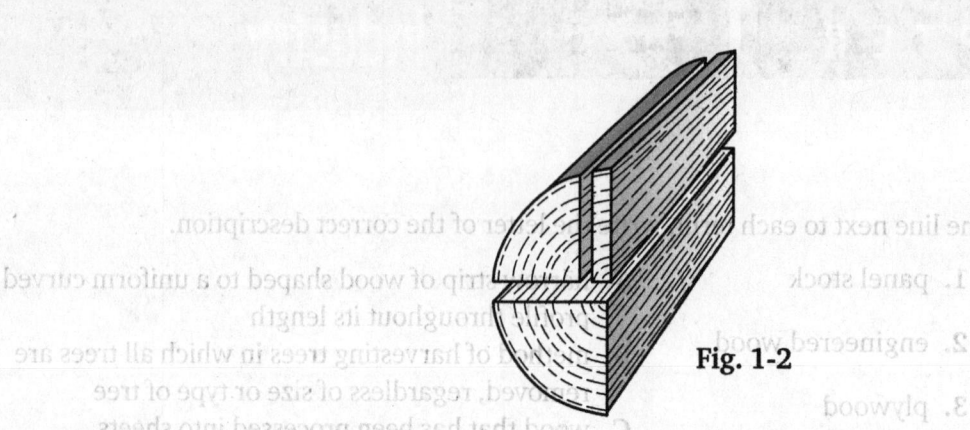

Fig. 1-2

_____ 16. Computer-controlled ___ are often used for cutting, drilling, and etching lumber.

_____ 17. ___ are often used for finishing operations and other processes that are too dangerous or tedious for humans.

_____ 18. The process of developing solutions to problems is known as the ___ process.

Matching

Directions: On the line next to each type of wood defect, write the letter of the correct description.

_____ 19. wane
_____ 20. warp
_____ 21. check
_____ 22. knot
_____ 23. stain
_____ 24. decay
_____ 25. split

A. discoloration on the surface of the wood
B. crook, cup, bow, or wind
C. dark round or oval spot where a branch had formed
D. rotting of the wood
E. slanting edge on a board created when the lumber was cut
F. lengthwise separation that looks like a large crack
G. separation, usually across the growth rings, that often appears at the end of a board

(Continued on next page)

8

Name _____ Date _____ Class _____

CHAPTER 2
Safety Practices

Matching

Directions: On the line next to each agency, write the letter of the correct description.

_____ 1. EPA
_____ 2. NIOSH
_____ 3. OSHA

A. approves dust masks and other personal protective equipment for use in the workshop
B. ensures that employers provide a safe working environment for all employees
C. establishes limits on the amounts of hazardous waste that can accumulate at a work site and for how long

Completion

Directions: On the line to the left of each sentence, write the word or phrase that correctly completes the sentence or answers the question.

_____ 4. Most accidents in the workshop can be avoided by developing a(n) ___ attitude.

_____ 5. A fire extinguisher that has the symbols shown in Fig. 2-1 is effective against fires involving combustibles such as paper and wood, flammable liquids, and ___ equipment.

Fig. 2-1

_____ 6. A fire extinguisher that has the symbols shown in Fig. 2-1 is known as a(n) ___ fire extinguisher.

_____ 7. A(n) ___ outlet or plug breaks the circuit when it detects a current leak to ground, protecting people from electrical shock.

_____ 8. For an electrical fire, you should remove the source of fuel by turning off the ___.

_____ 9. Workshops should be well ___ to keep fresh air circulating and prevent an accumulation of fumes.

_____ 10. When sanding, you should wear a dust mask or ___ that filters out harmful particles.

(Continued on next page)

Name _____ Date _____ Class _____

_____ 11. Employers are required to make ___ Safety Data Sheets available for hazardous materials in the workshop.

_____ 12. Fig. 2-2 shows three different kinds of ___.

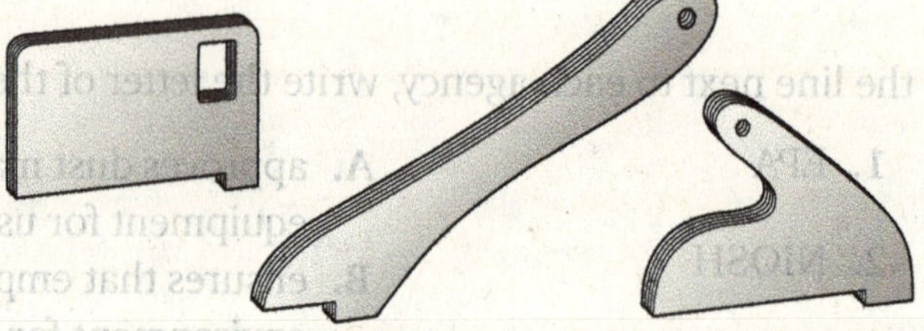

Fig. 2-2

_____ 13. South American mahogany and Western red cedar are two examples of woods that are ___.

_____ 14. Safety equipment that is designed to be worn is known as ___ protective equipment.

_____ 15. ___ are boards that help you press a workpiece against the fence or table of a stationary tool.

_____ 16. A(n) ___ shows a map of the building with at least two escape routes in case of fire. *(three words)*

True or False

Directions: Read each statement carefully. If the statement is true, write **True** in the blank to the left of that numbered item. If the statement is false, write **False** in the blank.

_____ 17. Safety goggles cannot be worn over prescription lenses.

_____ 18. Many power tools generate harmful levels of noise.

_____ 19. In an accident in which an object is embedded in the eye, never try to remove the object; call for medical help instead.

_____ 20. Most dangerous chemicals cannot enter the body by absorption through the skin.

(Continued on next page)

Name _____ Date _____ Class _____

CHAPTER 3
Designing and Planning

Matching

Directions: On the line next to each principle of design, write the letter of the correct description.

_____ 1. proportion
_____ 2. harmony
_____ 3. formal balance
_____ 4. emphasis
_____ 5. informal balance

A. design features appear to have equal weight, even though they are not exactly alike
B. the size relationship of parts or features
C. one half of an object appears to be the mirror image of the other half
D. stress or accent on one feature of the design
E. all of the parts, colors, shapes, and textures work well together

Completion

Directions: On the line to the left of each sentence, write the word or phrase that correctly completes the sentence or answers the question.

_____ 6. Three keys to good design are appearance, solid construction, and ___.

_____ 7. What type of drawing is shown in Fig. 3-1?

_____ 8. In an isometric drawing, the edges of the object form three equal angles of ___ degrees each.

_____ 9. Working drawings that show how a project looks when taken apart are called ___ drawings.

_____ 10. How many board feet are in a piece of stock that is 2 inches × 8 inches × 6 feet?

_____ 11. How many lineal feet are in a piece of stock that is 1 inch × 6 inches × 12 feet?

_____ 12. Lumber that is cut to a widely used size and shape is known as ___ stock.

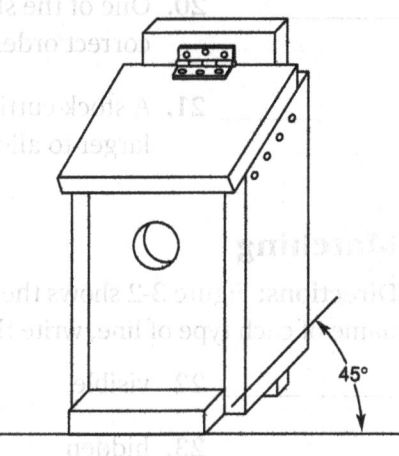

Fig. 3-1

(Continued on next page)

11

Name _____ Date _____ Class _____

_____ 13. A drawing in which a smaller measurement is used to represent a larger measurement is drawn to ___.

_____ 14. The process of measuring and marking stock to size and shape is called ___.

_____ 15. A(n) ___ drawing usually shows the front, top, and end views of an object.

True or False

Directions: Read each statement carefully. If the statement is true, write **True** in the blank to the left of that numbered item. If the statement is false, write **False** in the blank.

_____ 16. Pictorial drawings show only one view.

_____ 17. The first step in designing, planning, and completing a woodworking project is to make a working drawing of the project.

_____ 18. A bill of materials is a complete list of the tools and equipment you will need to complete a project.

_____ 19. To determine a dimension, it is a good idea to measure the drawing.

_____ 20. One of the steps in planning a project is to list the steps you will follow in the correct order.

_____ 21. A stock-cutting list is similar to a bill of materials, but the dimensions are slightly larger to allow for cutting and other operations.

Matching

Directions: Figure 3-2 shows the lines commonly used on working drawings. On the line next to the name of each type of line, write the letter of the illustration that shows that line type.

_____ 22. visible

_____ 23. hidden

_____ 24. centerline

_____ 25. dimension line

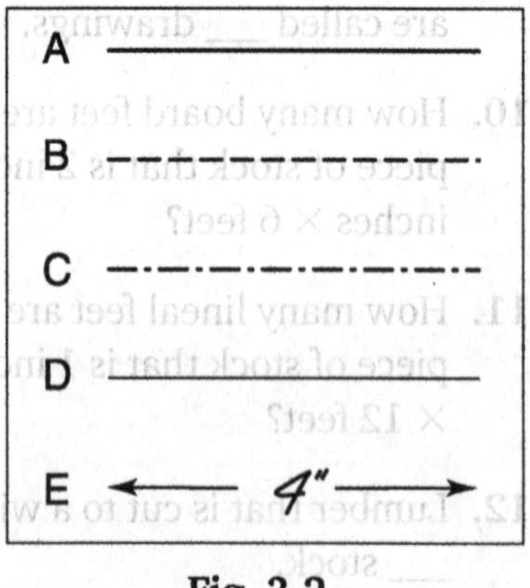

Fig. 3-2

12

Name _____ Date _____ Class _____

CHAPTER 4
Measuring and Cutting

Matching

Directions: On the line next to each type of cut, write the letter of the correct definition.

_____ 1. miter cut
_____ 2. bevel cut
_____ 3. crosscut
_____ 4. rip cut

A. angled cut made along the edge or end of the stock
B. cut made with the grain to cut stock to width
C. angled cut across the face of the stock
D. cut made across the wood grain to cut stock to length

Completion

Directions: On the line to the left of each sentence, write the word or phrase that correctly completes the sentence or answers the question.

_____ 5. The ___ system of measure is used most in the United States.

_____ 6. The stock shown in Fig. 4-1 is ___ centimeters long. (Give your answer to the nearest tenth of a centimeter.)

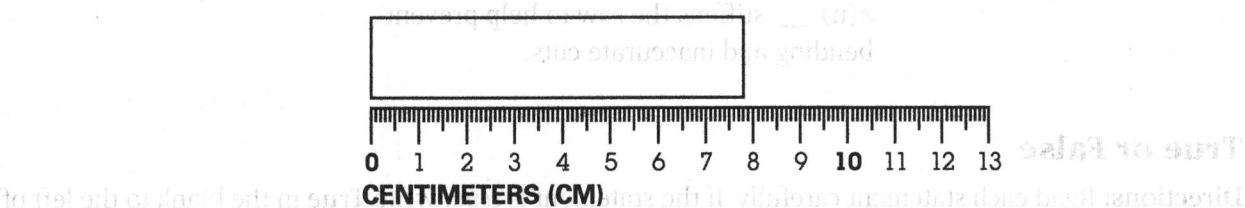

Fig. 4-1

_____ 7. What is the length of the stock shown in Fig. 4-1 in millimeters?

_____ 8. What is the length of the stock shown in Fig. 4-2? Give your answer to the nearest eighth of an inch.

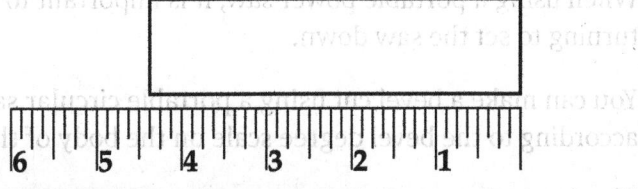

Fig. 4-2

(Continued on next page)

Name _____ Date _____ Class _____

_____ 9. The best tool to measure long lengths is the ___. *(two words)*

_____ 10. The two rules most often used to measure short lengths are the ___ rule and the bench rule.

_____ 11. To locate the center of a circle, the best tool to use is a(n) ___. *(two words)*

_____ 12. The adjustable blade in a sliding ___ allows you to lay out angles other than 90-degree angles.

_____ 13. Hold a rule across the edge to measure the ___ of the stock.

_____ 14. The most common marking tool for both rough and finished lumber is a(n) ___. *(two words)*

_____ 15. The back-and-forth (or up-and-down) motion of a(n) ___ saw makes rough cuts in wood, plaster, metal, and other materials.

_____ 16. The area shown in Fig. 4-3 is where the saw actually cuts. This area is known as the ___.

_____ 17. To make an accurate inside measurement, use a tape measure and a(n) ___ square.

_____ 18. The extra band of metal across the back of a(n) ___ stiffens the saw to help prevent bending and inaccurate cuts.

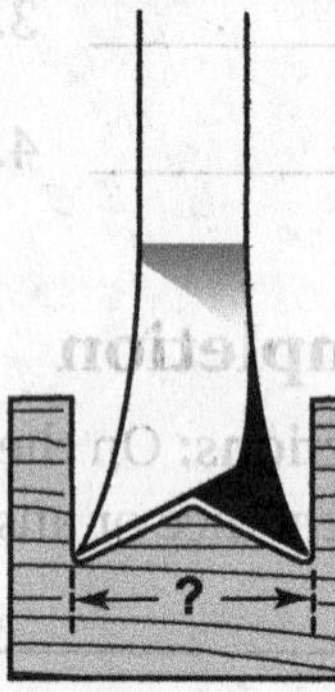

Fig. 4-3

True or False

Directions: Read each statement carefully. If the statement is true, write **True** in the blank to the left of that numbered item. If the statement is false, write **False** in the blank.

_____ 19. Before measuring the length of stock, use a try square to check its squareness.

_____ 20. Levels can be used to check both horizontal and vertical surfaces.

_____ 21. The ripsaw is most often used to cut stock with the grain.

_____ 22. When using a portable power saw, it is important to wait until the blade stops turning to set the saw down.

_____ 23. You can make a bevel cut using a portable circular saw by setting the blade according to the bevel degree scale on the body of the saw.

_____ 24. The dovetail saw is designed to cut a wide kerf for sawing curves and irregular shapes.

_____ 25. You should use a framing square to mark wide stock for cutting.

14

Name _____ Date _____ Class _____

CHAPTER 5
Nailing

Matching

Directions: Identify each part of the hammer shown in Fig. 5-1. On the line next to the name of each part, write the letter from the illustration that shows that part.

_____ 1. face

_____ 2. adze eye

_____ 3. handle

_____ 4. cheek

_____ 5. wedges

_____ 6. claw

_____ 7. neck

_____ 8. head

_____ 9. poll

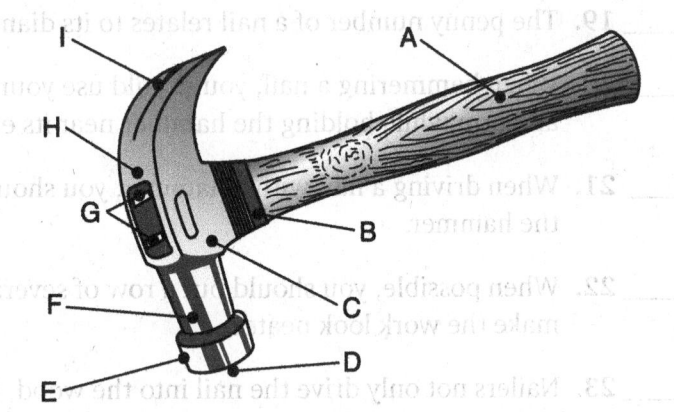

Fig. 5-1

Completion

Directions: On the line to the left of each sentence, write the word or phrase that correctly completes the sentence or answers the question.

_____ 10. A(n) ___ nail is a rather heavy nail with a small head.

_____ 11. The nails used in rough carpentry are known as ___ nails.

_____ 12. A(n) ___ is used to drive the head of a nail below the surface of wood. *(two words)*

_____ 13. Nailing the end of one piece of wood to the side of another piece by driving the nails into both sides at an angle is known as ___.

_____ 14. The nailers most often used in woodworking are finish and ___ nailers.

_____ 15. A(n) ___ has a nail slot on one end and a chisel-shape on the other. *(two words)*

(Continued on next page)

15

Name _____ Date _____ Class _____

_____ 16. The type of nail used in fine cabinet work is the ___ nail.

_____ 17. To remove a nail, force the ___ of a hammer under the head of the nail and pull on the handle.

True or False

Directions: Read each statement carefully. If the statement is true, write **True** in the blank to the left of that numbered item. If the statement is false, write **False** in the blank.

_____ 18. The face of a claw hammer should be a flat surface.

_____ 19. The penny number of a nail relates to its diameter.

_____ 20. When hammering a nail, you should use your wrist as well as your elbow and arm while holding the hammer near its end.

_____ 21. When driving a nail with a hammer, you should watch the head of the nail, not the hammer.

_____ 22. When possible, you should put a row of several nails along the same grain to make the work look neater.

_____ 23. Nailers not only drive the nail into the wood, but also set the nail.

_____ 24. It is better to use several light taps to drive a nail rather than a only a few heavier blows.

_____ 25. When you are using a hammer, you and everyone around you should wear safety glasses.

(Continued on next page)

16

Name _____ Date _____ Class _____

Chapter 6: Drilling

Matching

Directions: Identify each part of the power drill and bit shown in Fig. 6-1. On the line next to the name of each part, write the letter from the illustration that shows the part.

_____ 1. collar

_____ 2. jaws

_____ 3. bit shank

_____ 4. chuck key

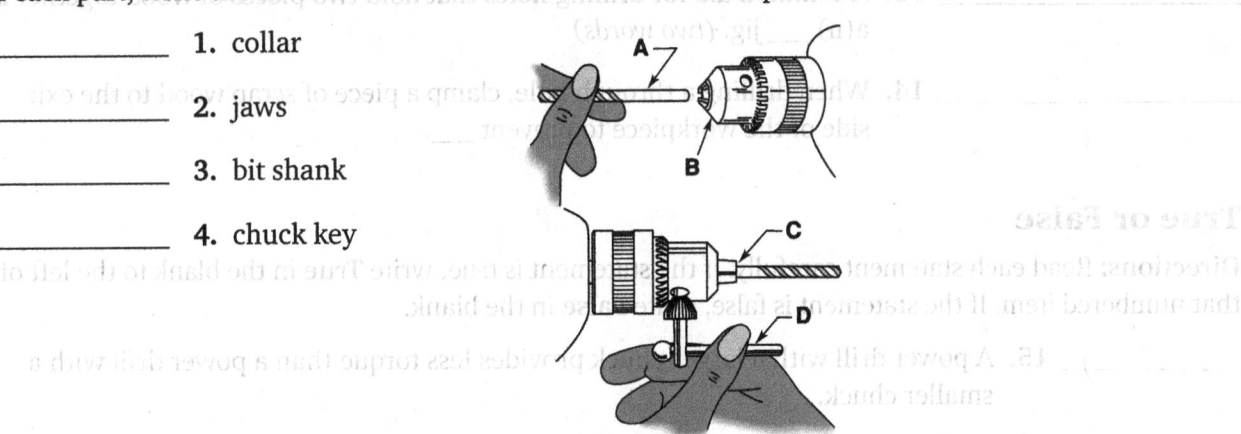

Fig. 6-1

Completion

Directions: On the line to the left of each sentence, write the word or phrase that correctly completes the sentence or answers the question.

_____ 5. The tool being used in Fig. 6-2 is a(n) ___. *(two words)*

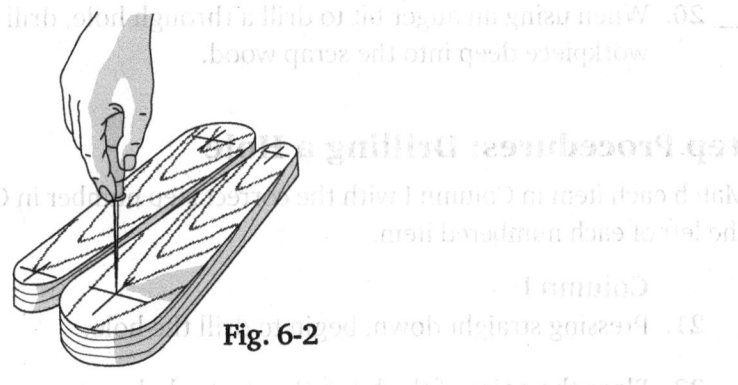

Fig. 6-2

_____ 6. The size of a twist drill is stamped on its ___.

_____ 7. A 5 stamped on an auger bit shows that the bit is ___ inch in diameter.

_____ 8. To use a twist drill, fasten it into the ___ of a bit brace.

(Continued on next page)

17

Name _____ Date _____ Class _____

_____ 9. A power drill for woodworking should be variable-speed and ___.

_____ 10. A(n) ___ can be used to drill holes ¼ inch or less in diameter by turning the crank. *(two words)*

_____ 11. The size of a brace is determined by the size of its ___.

_____ 12. A depth ___ controls the depth of the hole being drilled.

_____ 13. A common aid for drilling holes that hold two pieces of wood together is a(n) ___ jig. *(two words)*

_____ 14. When drilling a through hole, clamp a piece of scrap wood to the exit side of the workpiece to prevent ___.

True or False

Directions: Read each statement carefully. If the statement is true, write **True** in the blank to the left of that numbered item. If the statement is false, write **False** in the blank.

_____ 15. A power drill with a larger chuck provides less torque than a power drill with a smaller chuck.

_____ 16. Variable-speed power drills allow you to adjust speed by turning a thumbscrew.

_____ 17. As the torque of a power drill increases, drill speed decreases.

_____ 18. Pliers are needed to change a drill bit in a power drill that has a keyless chuck.

_____ 19. The only difference between a "power drill" and a "power screwdriver" is the type of bit that has been inserted into the chuck.

_____ 20. When using an auger bit to drill a through hole, drill all the way through the workpiece deep into the scrap wood.

Step-by-Step Procedures: Drilling a Hole

Directions: Match each item in Column I with the correct step number in Column II. Write one letter in the blank at the left of each numbered item.

Column I

_____ 21. Pressing straight down, begin to drill the hole.

_____ 22. Place the point of the bit in the starter hole.

_____ 23. Choose the correct size bit and fasten it to the drill.

_____ 24. Use a scratch awl to make a small hole at the center of-the-hole-location.

_____ 25. Clamp the workpiece in a vise or to the workbench.

Column II
A. Step 1
B. Step 2
C. Step 3
D. Step 4
E. Step 5

Name _____ Date _____ Class _____

CHAPTER 7
Planing, Chiseling, and Sanding

Matching

Directions: Identify each part of the block plane shown in Fig. 7-1. On the line next to the name of each part, write the letter from the illustration that shows that part.

_____ 1. adjustment nut

_____ 2. bottom

_____ 3. eccentric plate

_____ 4. cam

_____ 5. finger rest knob

_____ 6. lever cap

_____ 7. lateral adjustment lever

_____ 8. cutter

Fig. 7-1

Completion

Directions: On the line to the left of each sentence, write the word or phrase that correctly completes the sentence or answers the question.

_____ 9. A(n) ___ is a straightedge cutting tool used to shape and trim wood.

_____ 10. To make a very fine surface on open-grain wood such as oak, first plane the wood and then use a(n) ___ to remove the small imperfections left by the plane. *(two words)*

_____ 11. Smoothing wood by rubbing it with an abrasive is known as ___.

_____ 12. Garnet and aluminum oxide are two types of ___ that are used to sand wood.

_____ 13. The type of warp shown in Fig. 7-2 is a ___.

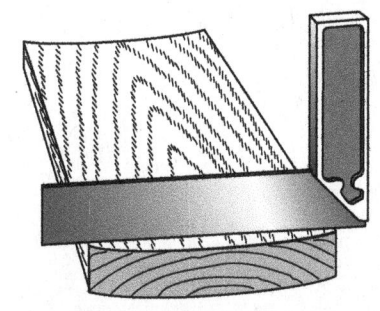

Fig. 7-2

(Continued on next page)

19

Name _____ Date _____ Class _____

_____ 14. The type of warp shown in Fig. 7-3 is a ___.

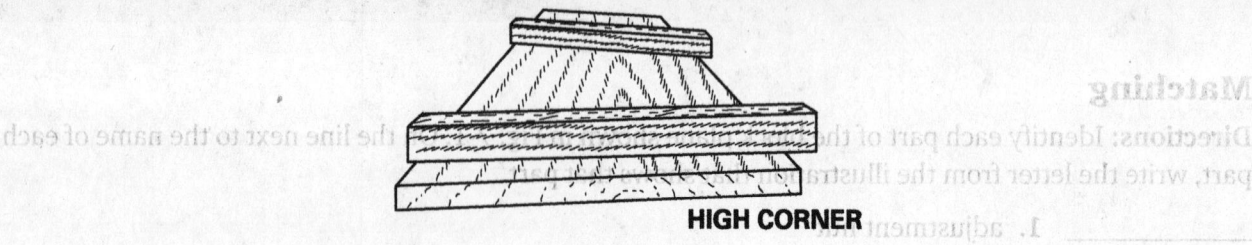

HIGH CORNER

HIGH CORNER

Fig. 7-3

True or False

Directions: Read each statement carefully. If the statement is true, write **True** in the blank to the left of that numbered item. If the statement is false, write **False** in the blank.

_____ 15. Most wood is surfaced on all four sides at the mill.

_____ 16. Always plane against the grain, never with it.

_____ 17. On the return stroke, the plane should be lifted.

_____ 18. When you plane the first edge, you should try to get the edge square with the face surface.

_____ 19. When planing a surface, apply pressure to the knob and handle as the plane starts to leave the wood surface.

_____ 20. When using a belt sander, you should let the rear of the moving belt touch the workpiece first.

(Continued on next page)

20

Name _____ Date _____ Class _____

CHAPTER 8: Butt, Biscuit, and Dowel Joints

Matching

Directions: Identify each of the joints shown in Fig. 8-1. On the line next to each joint name, write the letter of the illustration that shows the joint.

_____ 1. end butt joint

_____ 2. edge butt joint

_____ 3. biscuit joint

_____ 4. dowel joint

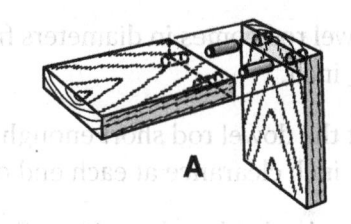

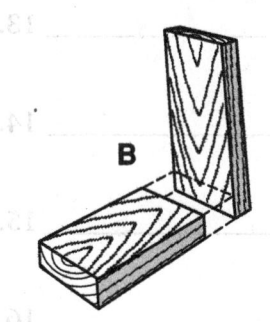

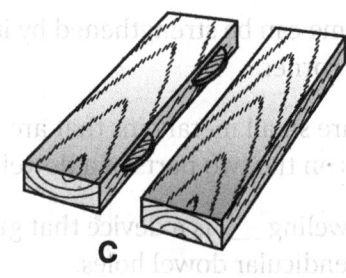

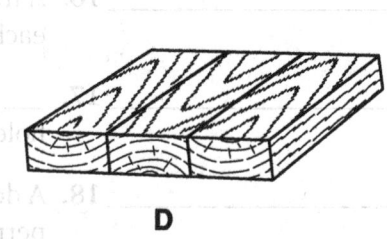

Fig. 8-1

True or False

Directions: Read each statement carefully. If the statement is true, write **True** in the blank to the left of that numbered item. If the statement is false, write **False** in the blank.

_____ 5. A biscuit joint does not require glue.

_____ 6. In making a butt joint, the butt ends must be square.

_____ 7. The butt joint is not an especially strong joint.

_____ 8. Screws grip better in cross grain than they do in end grain.

_____ 9. In making an edge dowel joint, choose dowel rod that is a little more than half the thickness of the stock.

_____ 10. A dowel rod should usually not be longer than 1 inch.

(Continued on next page)

21

Name _____ Date _____ Class _____

Completion

Directions: On the line to the left of each sentence, write the word or phrase that correctly completes the sentence or answers the question.

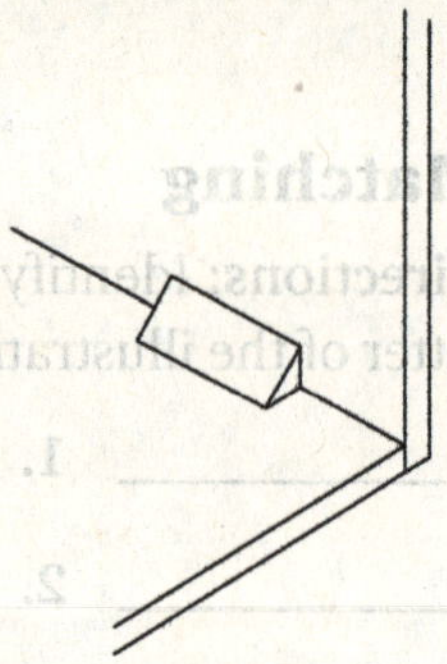

_____ 11. Figure 8-2 shows how a(n) ___ block can be used to strengthen a butt joint.

_____ 12. What type of wood is usually used to make dowel rod?

_____ 13. Dowel rod comes in diameters from ⅛ to ___ inch.

_____ 14. Cut the dowel rod short enough to allow about ___ inch clearance at each end of the joint.

Fig. 8-2

_____ 15. To make the dowels easier to fit into the holes, cut a slight ___ at each end.

_____ 16. A frame can be strengthened by installing one or more ___ at each corner.

_____ 17. ___ are small metal pins that are used to match and mark the location of holes on the two parts of a dowel joint. (two words)

_____ 18. A doweling ___ is a device that guides the drill to cut straight, perpendicular dowel holes.

Step-by-Step Procedures: Making a Biscuit Joint

Directions: Match each item in Column I with the correct step number in Column II. Write one letter in the blank at the left of each numbered item.

Column I

_____ 19. Clamp the two pieces of wood to be joined securely to the bench.

_____ 20. Clamp the two pieces together until the glue has cured.

_____ 21. Lay out the location of the joint on the adjoining wood surfaces.

_____ 22. Apply water-based glue to the slots and the surfaces.

_____ 23. Create the slots for the biscuits.

_____ 24. Insert the biscuits into the slots of the first piece and then fit the second piece to it, shifting them slightly for alignment if necessary.

_____ 25. Adjust the fences of the machine for vertical location of the slots and for depth of cut.

Column II
A. Step 1
B. Step 2
C. Step 3
D. Step 4
E. Step 5
F. Step 6
G. Step 7

(Continued on next page)

Name _____ Date _____ Class _____

CHAPTER 9

Rabbet Joint

Completion

Directions: On the line to the left of each sentence, write the word or phrase that correctly completes the sentence or answers the question.

_____ 1. The ___ of a rabbet should be from one-half to two-thirds the thickness of the stock.

_____ 2. The correct saw for cutting a rabbet by hand is the ___.

_____ 3. A piece of scrap stock is clamped over the ___ with hand screws to help ensure a square cut and to prevent the hand saw from damaging the wood. *(two words)*

_____ 4. When cutting a rabbet using a table saw, use the ___ as a guide. *(two words)*

_____ 5. To cut a rabbet using a router, use a(n) ___ bit.

_____ 6. To create a smooth rabbet joint, ___ off excess material after the glue has cured so both edges of the joint are flush.

True or False

Directions: Read each statement carefully. If the statement is true, write **True** in the blank to the left of that numbered item. If the statement is false, write **False** in the blank.

_____ 7. The rabbet joint is often used in drawer construction.

_____ 8. The back panels of cases and cabinets are often inset using a rabbet joint.

_____ 9. Fig. 9-1 shows the method to mark the depth of a rabbet cut.

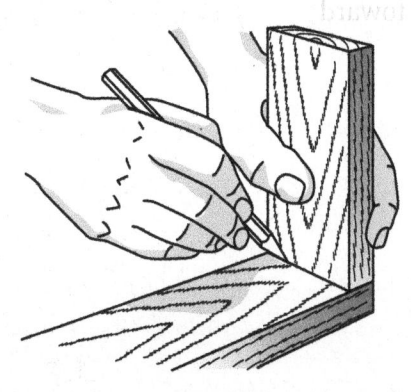

Fig. 9-1

(Continued on next page)

23

Name _____ Date _____ Class _____

_____ 10. When using a table saw to cut a rabbet, you should not allow your fingers to come closer than 5 inches to the saw blade.

_____ 11. When cutting a rabbet using hand tools, make the shoulder cut just inside the layout line.

_____ 12. After the shoulder cut has been made, a chisel or a saw can be used to remove excess stock.

_____ 13. Cutting a rabbet with a router requires two passes.

_____ 14. When removing excess stock using a table saw, the material to be removed should be facing out so that it can be safely separated from the workpiece.

_____ 15. The rabbet joint is strong enough that nails or screws are not usually needed to reinforce the glue.

Step-by-Step Procedures: Using a Table Saw to Cut a Rabbet

Directions: Match each item in Column I with the correct step number in Column II. Write one letter in the blank at the left of each numbered item.

Column I

_____ 16. Make the first cut.

_____ 17. Place the workpiece face up on the table with the edge against the fence.

_____ 18. Make the second cut.

_____ 19. Set the saw-blade projection to the width of the rabbet and the fence to the depth of the rabbet.

_____ 20. Place the workpiece on its end with the edge toward the saw blade and the face toward the fence.

Column II
A. Step 1
B. Step 2
C. Step 3
D. Step 4
E. Step 5

24

Name _____ Date _____ Class _____

Dado Joint
CHAPTER 10

Matching

Directions: Identify each of the joints shown in Fig. 10-1. On the line next to each joint name, write the letter of the illustration that shows the joint.

_____ 1. dado joint

_____ 2. blind dado joint

_____ 3. rabbet-and-dado joint

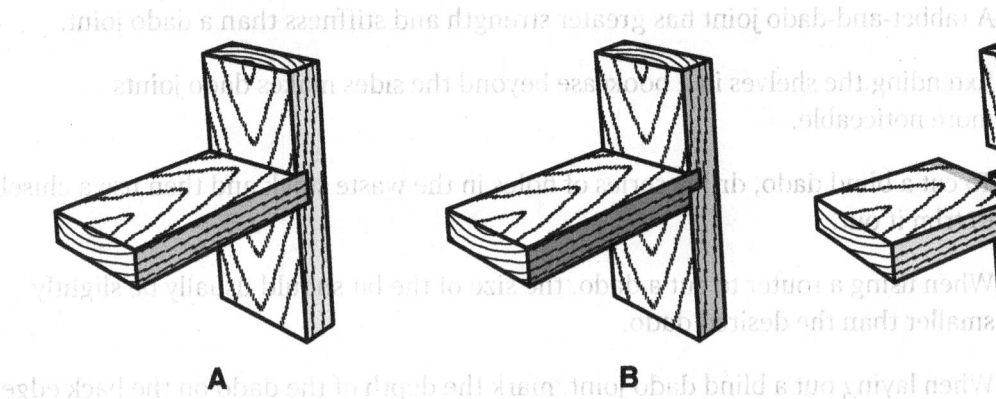

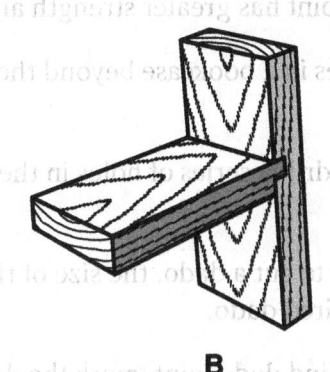

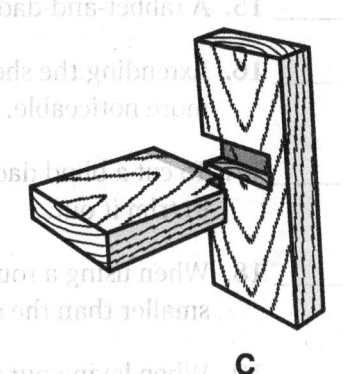

A   B   C

Fig. 10-1

Completion

Directions: On the line to the left of each sentence, write the word or phrase that correctly completes the sentence or answers the question.

_____ 4. The correct depth of a dado joint is ___ the thickness of the workpiece.

_____ 5. The dado is a groove cut across the ___ of the wood.

_____ 6. A dado joint is assembled in the same way as a(n) ___ joint.

_____ 7. One method of hiding a dado joint is to use face ___.

_____ 8. To lay out a blind dado joint, lay out the length of the dado from the back edge to within ___ to ¾ inch from the front edge.

_____ 9. The bit used with a portable router to cut a dado is a(n) ___ shank.

_____ 10. A table saw or a radial-arm saw equipped with a(n) ___ can make a dado quickly. *(two words)*

(Continued on next page)

Name _____ Date _____ Class _____

_____ 11. A dado is a strong joint because it provides a larger shared ___ area than most other joints and forms a supporting lip.

True or False

Directions: Read each statement carefully. If the statement is true, write **True** in the blank to the left of that numbered item. If the statement is false, write **False** in the blank.

_____ 12. A dado joint is a good choice for building a ladder.

_____ 13. If a dado joint has been cut too wide, plane the second piece slightly to make it fit the dado.

_____ 14. The saw kerf for cutting a dado joint should split the layout line.

_____ 15. A rabbet-and-dado joint has greater strength and stiffness than a dado joint.

_____ 16. Extending the shelves in a bookcase beyond the sides makes dado joints more noticeable.

_____ 17. To cut a blind dado, drill a series of holes in the waste stock and then use a chisel to trim it out.

_____ 18. When using a router to cut a dado, the size of the bit should usually be slightly smaller than the desired dado.

_____ 19. When laying out a blind dado joint, mark the depth of the dado on the back edge only.

_____ 20. If a dado joint is slightly too snug, you can usually force it to fit by tapping lightly with a mallet.

Step-by-Step Procedures: Making a Rabbet-and-Dado Joint

Directions: Match each item in Column I with the correct step number in Column II. Write one letter in the blank at the left of each numbered item.

	Column I	Column II
_____ 21.	Make the dado.	A. Step 1
_____ 22.	Lay out and cut the rabbet.	B. Step 2
_____ 23.	Fit the tongue of the rabbet into the dado.	C. Step 3
_____ 24.	Lay out the position of the dado joint by superimposing the tongue of the rabbet.	D. Step 4
_____ 25.	Mark the width of the dado.	E. Step 5

26

Name _____ Date _____ Class _____

CHAPTER 11
Lap Joint

Matching

Directions: Identify each type of lap joint shown in Fig. 11-1. On the line next to the name of each lap joint, write the letter from the illustration that shows that joint.

_____ 1. cross-lap

_____ 2. edge cross-lap

_____ 3. full-lap

_____ 4. half-lap

_____ 5. finger lap

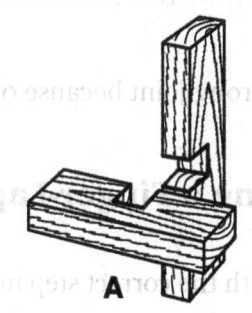

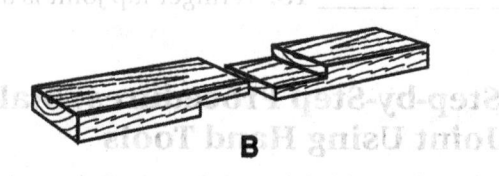

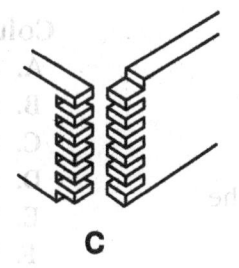

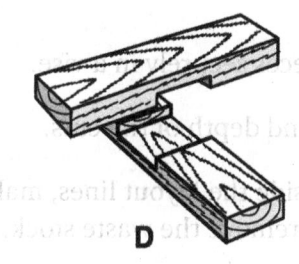

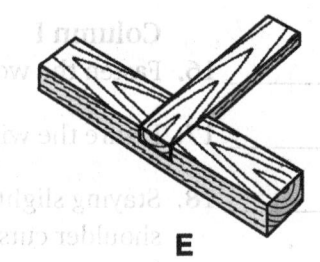

Fig. 11-1

Completion

Directions: On the line to the left of each sentence, write the word or phrase that correctly completes the sentence or answers the question.

_____ 6. A lap joint in which two pieces are joined without additional processing is known as a(n) ___-lap joint.

_____ 7. The most common lap joint is the ___-lap joint.

_____ 8. A(n) ___ should be used to trim out a lap joint.

_____ 9. A table saw with either a regular saw blade or a(n) ___ tool can be used to cut a lap joint.

_____ 10. When pieces to be joined are of different thicknesses, a(n) ___-lap joint is used.

_____ 11. In a(n) ___-lap joint, the pieces that are to be joined must have the same thickness.

(Continued on next page)

27

Name _____ Date _____ Class _____

True or False

Directions: Read each statement carefully. If the statement is true, write **True** in the blank to the left of that numbered item. If the statement is false, write **False** in the blank.

_____ 12. The greater the overlap in a lap joint, the stronger the joint will be.

_____ 13. In a cross-lap joint in the center of both pieces, one piece should be exactly twice the thickness of the other piece.

_____ 14. If the fit of a lap joint is too tight, you should plane a little from the edge of one piece.

_____ 15. A finger-lap joint is a strong joint because of its extensive glue area.

Step-by-Step Procedures: Making a Finger-Lap Joint Using Hand Tools

Directions: Match each item in Column I with the correct step number in Column II. Write one letter in the blank at the left of each numbered item.

Column I

_____ 16. Fasten the workpieces securely in a vise.

_____ 17. Figure the width and depth of the cuts.

_____ 18. Staying slightly inside the layout lines, make the shoulder cuts and remove the waste stock.

_____ 19. Place the pieces together, offsetting them by the width of one finger.

_____ 20. Select a drill bit that is the same width of the notches or slightly smaller.

_____ 21. Smooth the cuts using sandpaper wrapped around a small wood block.

_____ 22. Hold the drill at right angles to the surface and slowly drill holes at the bottom of the notches. Remove the wood block.

_____ 23. Square the bottom of the notches with a chisel and clean the corners carefully.

_____ 24. Lay out the cuts on the workpieces.

_____ 25. Clamp a block of scrap wood to the pieces behind the joint location to prevent splintering when the drill bit goes through the workpieces.

Column II
A. Step 1
B. Step 2
C. Step 3
D. Step 4
E. Step 5
F. Step 6
G. Step 7
H. Step 8
I. Step 9
J. Step 10

(Continued on next page)

28

Name _____ Date _____ Class _____

CHAPTER 12
Miter Joint

Matching

Directions: Calculate the angle of the cuts needed for miter joints in each of the picture frames shown in Fig. 12-1. On the line next to each angle, write the letter of the correct part.

_____ 1. 67.5 degrees

_____ 2. 60 degrees

_____ 3. 54 degrees

_____ 4. 45 degrees

_____ 5. 30 degrees

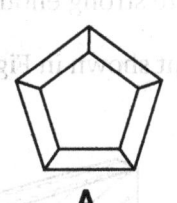

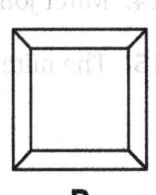

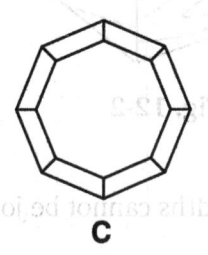

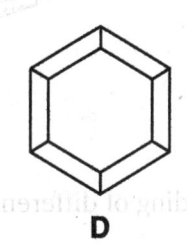

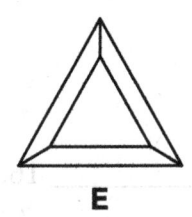

Fig. 12-1

Completion

Directions: On the line to the left of each sentence, write the word or phrase that correctly completes the sentence or answers the question.

_____ 6. The simple miter joint is usually cut at an angle of ___ degrees.

_____ 7. Miter boxes can be adjusted to cut angles from ___ to 90 degrees.

_____ 8. Four ways to strengthen a miter joint are to add ___, dowels, keys, or biscuits.

_____ 9. A miter joint can be either ___ or on edge.

_____ 10. ___ is very important in making the angle cuts for a miter joint because being even being a couple of degrees off can throw the project out of square.

_____ 11. A(n) ___ edge must be cut into the underside opening of a frame for the glass and picture.

(Continued on next page)

29

Name _____ Date _____ Class _____

_____ 12. To create your own decorative frame, you can choose from among hundreds of patterns of ___.

_____ 13. Before applying glue to a frame, use a try square to be sure the corners fit properly, and use a rule to measure the ___.

True or False

Directions: Read each statement carefully. If the statement is true, write **True** in the blank to the left of that numbered item. If the statement is false, write **False** in the blank.

_____ 14. Miter joints are strong enough that they rarely need to be reinforced.

_____ 15. The miter joint shown in Fig. 12-2 is on edge.

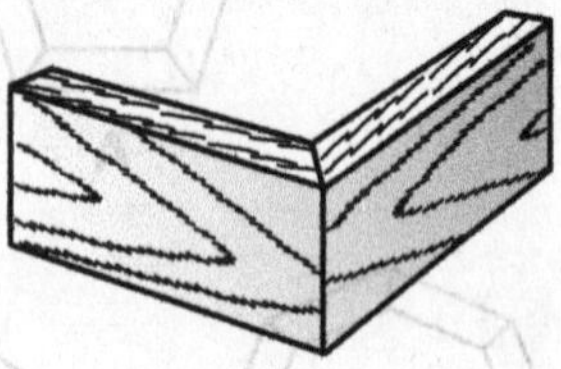

Fig. 12-2

_____ 16. Molding of different widths cannot be joined using a miter joint.

_____ 17. A miter joint can be cut with a homemade miter box.

_____ 18. To calculate the length to lay out for the sides of a frame, add twice the width of the framing material (from shoulder of the rabbet to the outside edge of the stock) to the length of the glass or picture to be framed.

_____ 19. Accuracy is not important in making angle cuts for a miter joint because the end grain of both pieces is hidden.

_____ 20. A miter joint must be clamped for nailing.

Name _____ Date _____ Class _____

CHAPTER 13
Mortise-and-Tenon Joint

Matching

Directions: Identify each type of mortise-and-tenon joint shown in Fig. 13-1. On the line next to the name of each type of joint, write the letter from the illustration that shows that joint.

_____ 1. open

_____ 2. round

_____ 3. blind

_____ 4. through

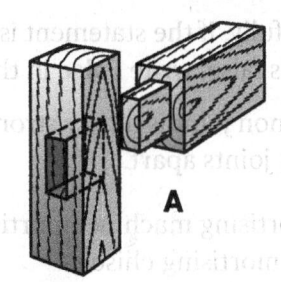

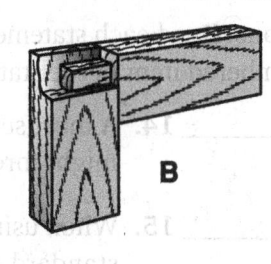

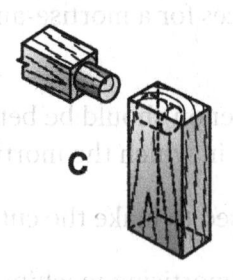

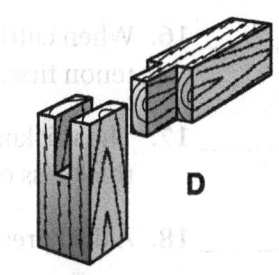

Fig. 13-1

Completion

Directions: On the line to the left of each sentence, write the word or phrase that correctly completes the sentence or answers the question.

_____ 5. The ___ of a mortise is the same as the thickness of the tenon.

_____ 6. The ___ of a mortise depends on whether the mortise is a blind or through mortise.

_____ 7. Before you apply glue to a mortise-and-tenon joint, the area must be-___.

_____ 8. A(n) ___ mortise is one that is cut all the way through the piece of wood.

_____ 9. The projecting piece of wood shaped to fit into the hole in a mortise-and-tenon joint is the ___.

_____ 10. When a(n) ___ mortise is used, the end of the tenon cannot be seen after the parts are assembled.

(Continued on next page)

Name _____ Date _____ Class _____

_____ 11. The width of a tenon should be ___ inches or less.

_____ 12. When cutting a mortise using hand tools, you can use a(n) ___ chisel to remove the waste stock.

_____ 13. Cut a small ___ around the end of a tenon to help it slip easily into the mortise.

True or False

Directions: Read each statement carefully. If the statement is true, write **True** in the blank to the left of that numbered item. If the statement is false, write **False** in the blank.

_____ 14. A mortise-and-tenon joint is a very strong joint that can withstand pressures that might force other joints apart.

_____ 15. When using a mortising machine, mortises should be sized according to the standard sizes of mortising chisels.

_____ 16. When cutting the pieces for a mortise-and-tenon joint, you should plan to cut the tenon first.

_____ 17. The thickness of the tenon should be between one-third and one-half the thickness of the piece in which the mortise is cut.

_____ 18. A drill press can be used to make the cuts for a mortise-and-tenon joint.

_____ 19. The speed at which a mortising machine is run depends on the size of the chisel and the wood's density.

_____ 20. The tenon should fit into the mortise loosely to allow room for gluing.

32

Name _____ Date _____ Class _____

CHAPTER 14
Dovetail Joints and Casework

Completion

Directions: On the line to the left of each sentence, write the word or phrase that correctly completes the sentence or answers the question.

_____ 1. The two pieces that are joined in a dovetail joint are the ___ piece and the tail piece.

_____ 2. When setting up the jig to cut a dovetail joint, you should fasten the tail piece to the ___ of the jig.

_____ 3. The joint shown in Fig. 14-1 is a(n) ___ dovetail joint.

_____ 4. You can make a dovetail joint using a dovetail jig and a router with a(n) ___ bit.

_____ 5. ___ is a box turned on its end or edge.

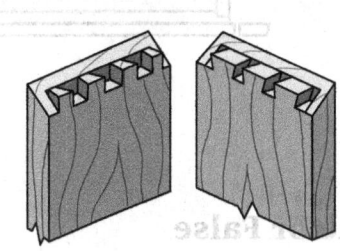

Fig. 14-1

_____ 6. The back of a case is usually fitted using a(n) ___ joint.

_____ 7. Table corners can be strengthened by using corner ___ such as the one shown in Fig. 14-2.

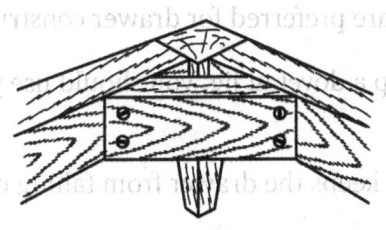

Fig. 14-2

_____ 8. To allow for shrinking and swelling, the drawer bottom should be cut slightly ___ than the width between the grooves.

_____ 9. To keep a drawer from sticking, rub the tight spot with ___.

_____ 10. Slotted strips called shelf ___ can be used with brackets or clips to support adjustable shelves in bookcases.

_____ 11. The vertical side pieces of a paneled door are known as ___.

(Continued on next page)

Name _____ Date _____ Class _____

_____ 12. ___ doors are used in places in which it is difficult to have doors that swing open and shut.

_____ 13. The horizontal parts that join the legs of a table or stool are called ___.

_____ 14. The drawer in Fig. 14-3A is a(n) ___ drawer.

_____ 15. The drawer in Fig. 14-3B is a(n) ___ drawer.

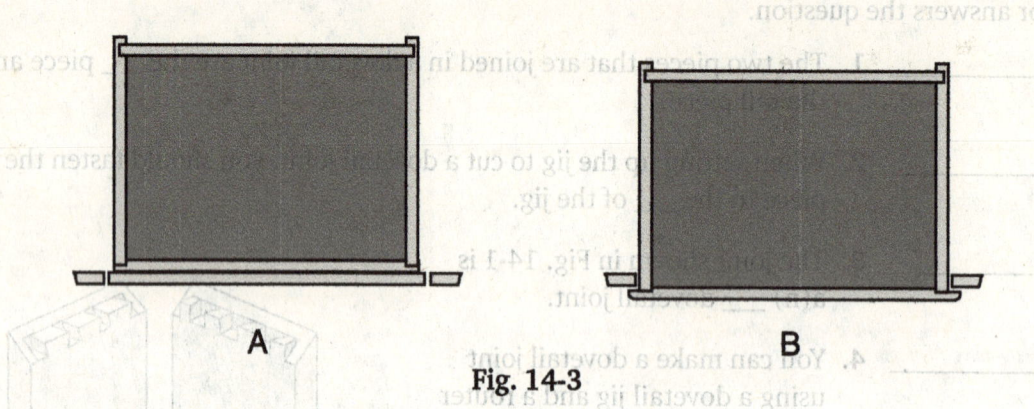

Fig. 14-3

True or False

Directions: Read each statement carefully. If the statement is true, write **True** in the blank to the left of that numbered item. If the statement is false, write **False** in the blank.

_____ 16. The bottom of a drawer should be glued in place.

_____ 17. A drawer guide must always be made of plastic so it slides easily.

_____ 18. Dovetail joints are preferred for drawer construction because of their strength.

_____ 19. When setting up a dovetail jig, you should use your actual workpiece to make the adjustments.

_____ 20. A drawer guide keeps the drawer from falling out of the opening.

34

CHAPTER 15 — Using Fasteners

Matching

Directions: Identify each type of fastener shown in Fig. 15-1. On the line next to the name of each item, write the letter from the illustration that shows the item.

_____ 1. finishing washer

_____ 2. drywall screw

_____ 3. oval head screw

_____ 4. flat washer

_____ 5. roundhead screw

_____ 6. lag screw

_____ 7. flathead screw

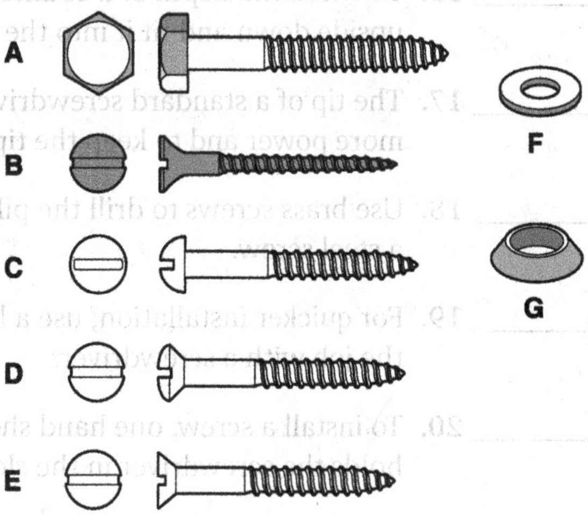

Fig. 15-1

Completion

Directions: On the line to the left of each sentence, write the word or phrase that correctly completes the sentence or answers the question.

_____ 8. The two types of holes that must be predrilled into wood before installing screws are the shank clearance hole and the ___ hole.

_____ 9. A shank clearance hole should be drilled the same size or slightly larger than the ___ of the screw.

_____ 10. A(n) ___ screwdriver is used to install or remove screws located in tight places where a standard screwdriver won't fit.

_____ 11. Large screws can be installed easily using a power drill with a(n) ___ bit.

_____ 12. Enlarging the top portion of a hole to a cone shape so that the head of a screw will be flush with the surface of the wood is called ___.

_____ 13. The ___ counterbore drills the shank clearance hole and pilot hole, countersinks, and drills plug holes for wooden plugs.

(Continued on next page)

Name _____ Date _____ Class _____

_____ 14. A standard screwdriver is used to install ___ screws.

_____ 15. A(n) ___-head screwdriver is made for driving crossed-head screws.

True or False

Directions: Read each statement carefully. If the statement is true, write **True** in the blank to the left of that numbered item. If the statement is false, write **False** in the blank.

_____ 16. To check the depth of a countersunk hole for a flathead screw, turn the screw upside down and fit it into the hole.

_____ 17. The tip of a standard screwdriver should be wider than the screw head to provide more power and to keep the tip from slipping from the screw.

_____ 18. Use brass screws to drill the pilot hole in hardwood to avoid shearing the head off a steel screw.

_____ 19. For quicker installation, use a hammer to install a screw part way, and then finish the job with a screwdriver.

_____ 20. To install a screw, one hand should turn the screwdriver while the other hand holds the screwdriver in the slot of the screw.

Matching

Directions: Identify each part of the screwdriver shown in Fig. 15-2. On the line next to the name of each item, write the letter from the illustration that shows the item.

_____ 21. head

_____ 22. blade

_____ 23. ferrule

_____ 24. tip

_____ 25. handle

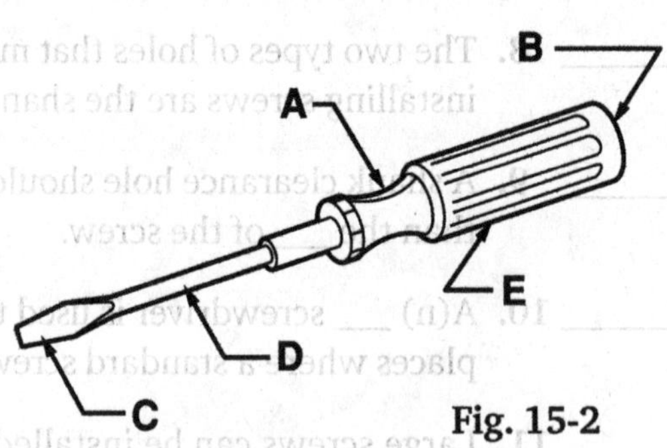

Fig. 15-2

(Continued on next page)

36

Name _____ Date _____ Class _____

CHAPTER 17
Installing Hardware

Matching

Directions: On the line next to each term, write the letter of the correct description.

_____ 1. cabinet hardware

_____ 2. structural hardware

_____ 3. catch

_____ 4. gain

_____ 5. hinge

_____ 6. repair plate

A. a hardware device for holding a door closed
B. a recess cut in a door and frame for a hinge
C. hardware that strengthens joints and holds unseen parts together
D. used to fix or strengthen various parts of a project
E. a piece of hardware that is used as a joint
F. hardware that makes projects with drawers and doors more usable

Completion

Directions: On the line to the left of each sentence, write the word or phrase that correctly completes the sentence or answers the question.

_____ 7. The best type of hinge to use when you do not want even the pivot point of the hinge to be visible is the ___ hinge.

_____ 8. The hinge shown in Fig. 17-1 is a(n) ___ hinge.

_____ 9. A(n) ___ butt hinge is one on which one or both of the leaves are shaped at the knuckle to allow for a closer fit.

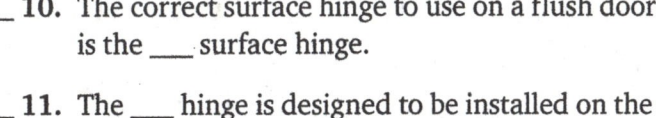

_____ 10. The correct surface hinge to use on a flush door is the ___ surface hinge.

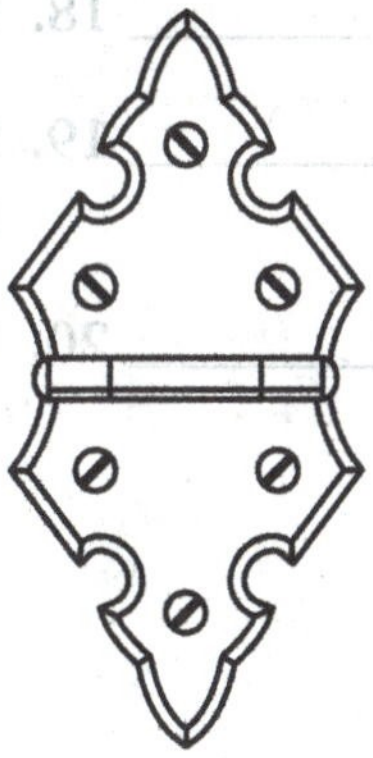

_____ 11. The ___ hinge is designed to be installed on the top and bottom of an overlay door.

_____ 12. When a(n) ___ hinge is used, only the pivot point can be seen when the door is closed.

_____ 13. The first step in installing drawer hardware is to ___ and mark the location for the hardware on the drawer.

_____ 14. The best way to support most drawers is to use a metal ___ guide set.

Fig. 17-1

(Continued on next page)

39

Name _____ Date _____ Class _____

True or False

Directions: Read each statement carefully. If the statement is true, write **True** in the blank to the left of that numbered item. If the statement is false, write **False** in the blank.

_____ 15. The type of repair plate shown in Fig. 17-2 is a flat corner plate.

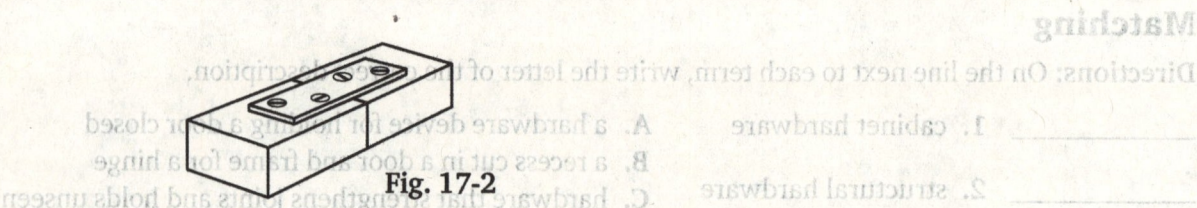

Fig. 17-2

_____ 16. Cabinet hardware should be chosen to match the style of a project.

_____ 17. The hinge shown in Fig. 17-3 is a half-swaged hinge.

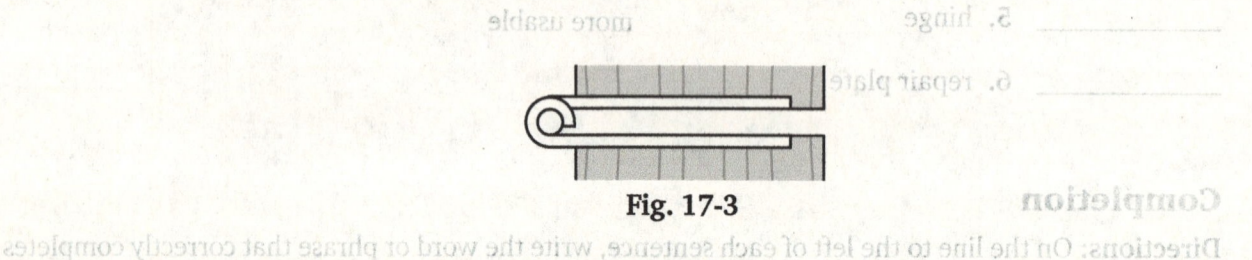

Fig. 17-3

_____ 18. Surface hinges usually require a gain.

_____ 19. Side guides are preferred over bottom guides for drawers because they can carry more weight and operate more smoothly.

_____ 20. T-plates are often used to hang cabinets and shelves.

(Continued on next page)

40

Name _____ Date _____ Class _____

Plastic Laminates

CHAPTER 18

Matching

Directions: Identify the parts of the laminated countertop in Fig. 18-1. On the line next to the name of each part, write the letter from the illustration that shows the part.

_____ 1. build-up strip

_____ 2. laminate

_____ 3. molding

_____ 4. substrate

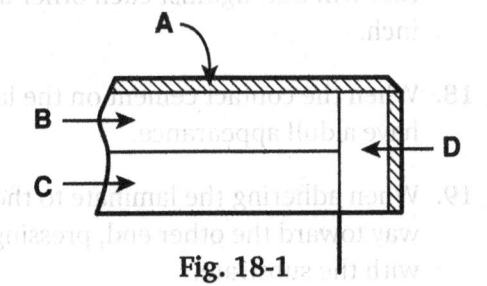

Fig. 18-1

Completion

Directions: On the line to the left of each sentence, write the word or phrase that correctly completes the sentence or answers the question.

_____ 5. Plastic laminate is made from layers of ___ paper.

_____ 6. High heat and ___ are used to produce the final 1/16-inch sheets of laminate.

_____ 7. The layers of paper that make up a plastic laminate are filled with ___.

_____ 8. The substrate for a plastic laminate is usually plywood, MDF, or ___.

_____ 9. Plastic laminate comes in sheets of up to 5 feet by ___ feet.

_____ 10. ___-tipped tools are preferred for cutting laminates because they remain sharp longer.

_____ 11. To cut laminate to rough length and width, use a router with a(n) ___ bit.

_____ 12. The adhesive used to glue laminate to a substrate is ___. *(two words)*

_____ 13. The layer to which the laminate is applied is known as the ___.

_____ 14. To provide a finished appearance, install ___ on the facing edges of objects that have been topped with plastic laminates.

(Continued on next page)

41

Name _____ Date _____ Class _____

True or False

Directions: Read each statement carefully. If the statement is true, write **True** in the blank to the left of that numbered item. If the statement is false, write **False** in the blank.

_____ 15. Plastic laminate can be cleaned with soap and water.

_____ 16. Eye protection is not necessary when using a router to cut plastic laminate.

_____ 17. When a surface must be made of more than one piece of laminate, cut edges that will butt against each other at the same time, overlapping the ends about 1 inch.

_____ 18. When the contact cement on the laminate is ready for bonding, the dry coats will have a dull appearance.

_____ 19. When adhering the laminate to the substrate, begin at one end and work your way toward the other end, pressing down on the laminate so that contact is made with the substrate.

_____ 20. It is not necessary to apply pressure to the laminate and substrate after applying the adhesive.

(Continued on next page)

42

Name _____ Date _____ Class _____

CHAPTER 19
Veneering

Matching

Directions: On the line next to each term, write the letter of the correct description.

_____ 1. veneering
_____ 2. flitch
_____ 3. substrate
_____ 4. plain edging
_____ 5. heat-adhesive edging
_____ 6. pressure-adhesive edging

A. a bundle of the slices from one log
B. edging that is simply attached with an adhesive such as white glue
C. the wood material to which veneer is applied
D. the process of applying a thin layer of fine wood to the surface of a wood of lesser quality
E. edging that is applied by peeling back the cover paper and fastening the edging in place
F. adhesive-backed edging that can be applied with hand pressure, followed by heat to activate the adhesive

Completion

Directions: On the line to the left of each sentence, write the word or phrase that correctly completes the sentence or answers the question.

_____ 7. Fig. 19-1 shows the ___ method of cutting veneer.

_____ 8. Fig. 19-2 shows the ___ method of cutting veneer.

_____ 9. Most veneers are ___ inch thick.

_____ 10. A(n) ___ saw, craft knife, or utility knife can be used to cut veneer.

_____ 11. Serious woodworkers often use a(n) ___ to apply pressure to veneer during the gluing process. *(two words)*

_____ 12. Veneer can be applied to a substrate using white or yellow glue or ___. *(two words)*

_____ 13. On some veneers, a(n) ___ provides the heat necessary to bond the adhesive to the substrate and veneer.

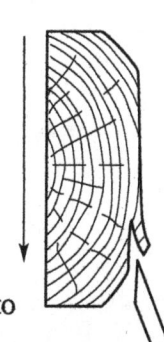

Fig. 19-1

Fig. 19-2

(Continued on next page)

43

Name _____ Date _____ Class _____

True or False

Directions: Read each statement carefully. If the statement is true, write **True** in the blank to the left of that numbered item. If the statement is false, write **False** in the blank.

_____ 14. One of the most important things about using veneer instead of solid wood is that good wood is used more efficiently.

_____ 15. Flat-cut veneers are generally less attractive than rotary-cut veneers because the grain pattern is spread out and irregular.

_____ 16. A surface that has been coated with contact cement will not adhere to any other surface that has not been coated with contact cement.

Matching

Directions: Identify the type of veneer matching used in each of the illustrations in Fig. 19-3. On the line next to each type of matching, write the letter from the illustration that shows that type.

_____ 17. diamond quartered

_____ 18. bookmatched

_____ 19. slipmatched

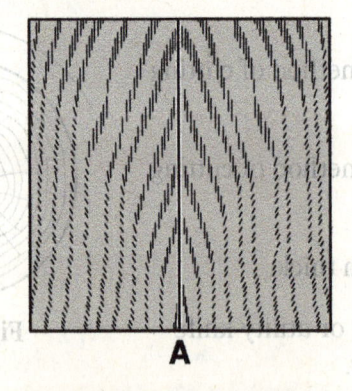

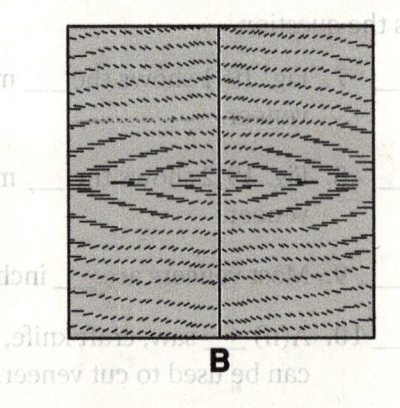

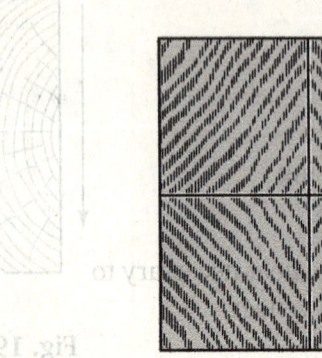

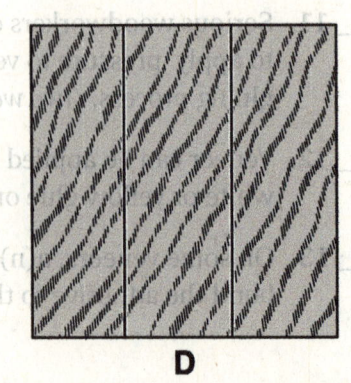

_____ 20. buttmatched

Fig. 19-3

(Continued on next page)

44

Name _____ Date _____ Class _____

Band Saw
CHAPTER 24

Matching

Directions: Identify the parts of the band saw shown in Fig. 24-1. On the line next to the name of each part, write the letter from the illustration that shows the part.

_____ 1. upper wheel guard

_____ 2. blade

_____ 3. table

_____ 4. guide post lock

_____ 5. guide post

_____ 6. switch

_____ 7. base

_____ 8. arm

_____ 9. lower wheel guard

_____ 10. blade guides

Fig. 24-1

Completion

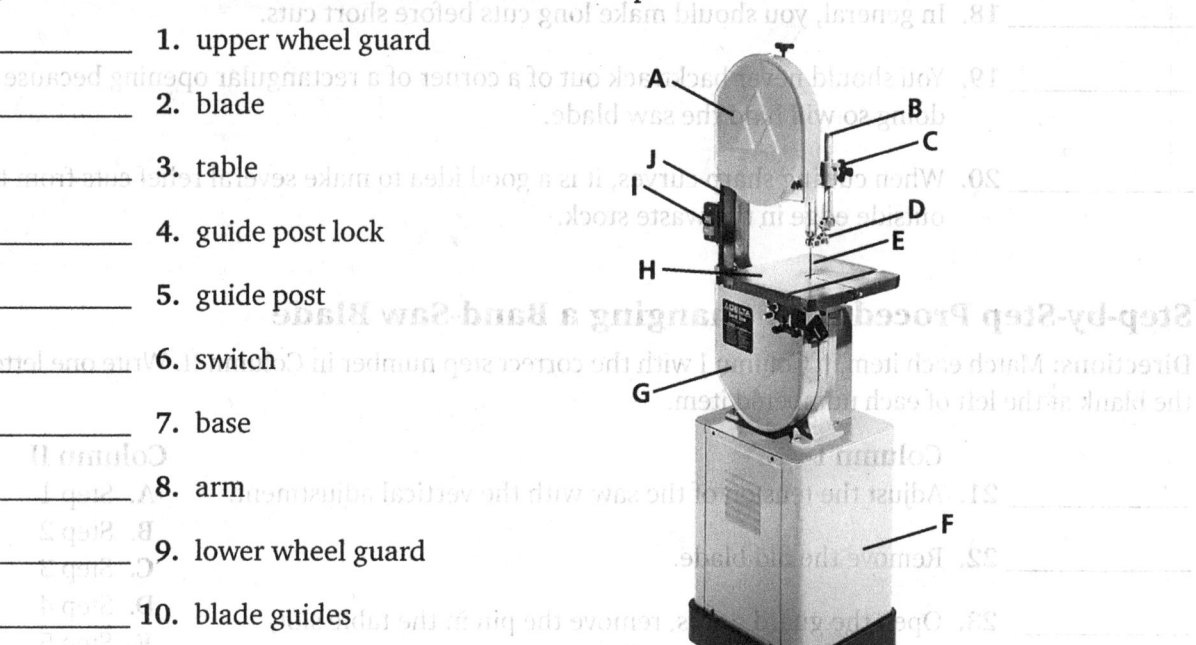

Directions: On the line to the left of each sentence, write the word or phrase that correctly completes the sentence or answers the question.

_____ 11. When using a band saw, you should maintain a(n) ___-inch margin of safety.

_____ 12. Sawing stock to reduce its thickness is known as ___.

_____ 13. The process of cutting several pieces at one time is known as ___. *(two words)*

_____ 14. To crosscut using a band saw, you should use a(n) ___ gauge.

_____ 15. The best way to cut a circle using a band saw is to use a(n) ___ jig.

(Continued on next page)

53

Name _____ Date _____ Class _____

True or False

Directions: Read each statement carefully. If the statement is true, write **True** in the blank to the left of that numbered item. If the statement is false, write **False** in the blank.

_____ 16. When operating a band saw, you should stand facing the blade and slightly to the left.

_____ 17. Move the stock into the blade as slowly as possible to avoid burning the wood.

_____ 18. In general, you should make long cuts before short cuts.

_____ 19. You should never backtrack out of a corner of a rectangular opening because doing so will bind the saw blade.

_____ 20. When cutting sharp curves, it is a good idea to make several relief cuts from the outside edge in the waste stock.

Step-by-Step Procedures: Changing a Band-Saw Blade

Directions: Match each item in Column I with the correct step number in Column II. Write one letter in the blank at the left of each numbered item.

Column I

_____ 21. Adjust the tension of the saw with the vertical adjustment.

_____ 22. Remove the old blade.

_____ 23. Open the guard doors, remove the pin in the table slot, and loosen the vertical adjustment.

_____ 24. Slip the new blade onto the upper and lower wheels, with the teeth pointing downward.

_____ 25. Disconnect the machine.

Column II
A. Step 1
B. Step 2
C. Step 3
D. Step 4
E. Step 5

54

Name _____ Date _____ Class _____

CHAPTER 25
Sliding Compound Miter Saw

Matching

Directions: Identify the parts of the compound miter saw shown in Fig. 25-1. On the line next to the name of each part, write the letter from the illustration that shows the part.

_____ 1. miter gauge

_____ 2. turntable

_____ 3. dust bag

_____ 4. miter lock handle

_____ 5. bevel gauge

_____ 6. saw guard

_____ 7. stock clamp

_____ 8. fence

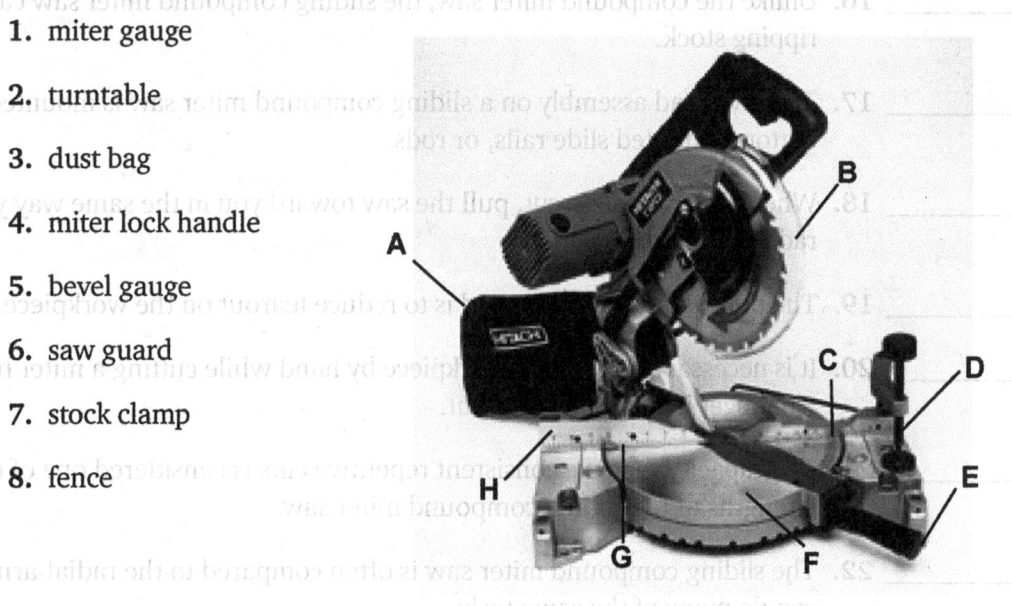

Fig. 25-1

Completion

Directions: On the line to the left of each sentence, write the word or phrase that correctly completes the sentence or answers the question.

_____ 9. The bevel angle on a sliding compound miter saw can be set at any angle from 0 to ___ degrees.

_____ 10. To set the angle of a miter, position the ___ according to a scale.

_____ 11. The plastic or fiberboard piece located in the center of the turntable that minimizes the gap between the side of the blade and the turntable is a(n) ___. (*two words*)

_____ 12. To operate a sliding compound miter saw, you must press both the ___ button and the power switch.

_____ 13. Workpieces should be tightly secured using the ___ clamps that come with the saw.

_____ 14. A cut that includes both a bevel and a miter is said to be cut at a(n) ___ angle.

(*Continued on next page*)

55

Name _____ Date _____ Class _____

_____ 15. To make a straight crosscut, both scales on the sliding compound miter saw should be set at ___ degrees.

True or False

Directions: Read each statement carefully. If the statement is true, write **True** in the blank to the left of that numbered item. If the statement is false, write **False** in the blank.

_____ 16. Unlike the compound miter saw, the sliding compound miter saw can be used for ripping stock.

_____ 17. The saw head assembly on a sliding compound miter saw is mounted on top- or bottom-mounted slide rails, or rods.

_____ 18. When making a slide cut, pull the saw toward you in the same way you would a radial-arm saw.

_____ 19. The purpose of the kerf board is to reduce tearout on the workpiece.

_____ 20. It is necessary to hold the workpiece by hand while cutting a miter because clamps would get in the way of the cut.

_____ 21. The capacity to make consistent repetitive cuts is considered one of the true strengths of the sliding compound miter saw.

_____ 22. The sliding compound miter saw is often compared to the radial-arm saw because it can do many of the same tasks.

_____ 23. On a sliding compound miter saw, the motion of the saw, not the settings, determines the angles of the cuts.

_____ 24. The faster you cut with a sliding compound miter saw, the smoother the cut will be.

_____ 25. When you finish a cut, you should release the power switch and allow the blade to coast to a full stop before lifting the saw.

(Continued on next page)

56

Name _____ Date _____ Class _____

CHAPTER 26 — Scroll Saw

Matching

Directions: Identify the parts of the scroll saw shown in Fig. 26-1. On the line next to the name of each part, write the letter from the illustration that shows the part.

_____ 1. saw stand

_____ 2. blade holder

_____ 3. flexible blow-on

_____ 4. blade tensioning

_____ 5. blade holder

_____ 6. overarm

_____ 7. electronic variable speed control

_____ 8. base

_____ 9. table tilt scale

_____ 10. hold-down

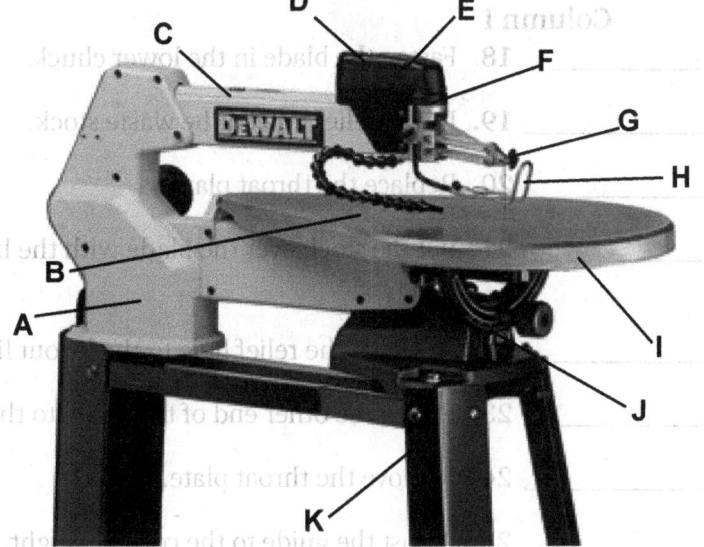

Fig. 26-1

Completion

Directions: On the line to the left of each sentence, write the word or phrase that correctly completes the sentence or answers the question.

_____ 11. The driving mechanism of a scroll saw converts rotating motion into up-and-down, or ___, motion.

_____ 12. The distance between the blade and the base of the overarm on a scroll saw is the ___ depth.

_____ 13. You should not allow your fingers to come any closer than ___ inches to the blade when cutting stock with a scroll saw.

_____ 14. When making intricate internal cuts, you should first drill a(n) ___ hole in the center of the waste stock.

_____ 15. ___, or marquetry, is a way of forming a design using contrasting kinds of wood.

(Continued on next page)

57

Name _____ Date _____ Class _____

_____ 16. When choosing a scroll saw blade, you should take both the material to be cut and the ___ of the pattern into consideration.

_____ 17. In general, you should choose the widest and ___ blade possible that will still allow you to follow the most complex details of your pattern.

Step-by-Step Procedures: Cutting an Internal Curve

Directions: Match each item in Column I with the correct step number in Column II. Write one letter in the blank at the left of each numbered item.

Column I

_____ 18. Fasten the blade in the lower chuck.

_____ 19. Drill a relief hole in the waste stock.

_____ 20. Replace the throat plate.

_____ 21. Put the stock over the blade with the blade going through the relief hole.

_____ 22. Cut from the relief hole to the layout line.

_____ 23. Fasten the other end of the blade to the upper chuck.

_____ 24. Remove the throat plate.

_____ 25. Adjust the guide to the correct height.

Column II

A. Step 1
B. Step 2
C. Step 3
D. Step 4
E. Step 5
F. Step 6
G. Step 7
H. Step 8

58

Name _____ Date _____ Class _____

CHAPTER 27 — Drill Press

Matching

Directions: Identify the parts of the drill press shown in Fig. 27-1. On the line next to the name of each part, write the letter from the illustration that shows the part.

_____ 1. on/off switch

_____ 2. chuck

_____ 3. quill feed lever

_____ 4. column

_____ 5. belt-and-pulley housing

_____ 6. tilting table

_____ 7. depth stop

_____ 8. motor

_____ 9. rack-and-pinion mechanism

_____ 10. spindle

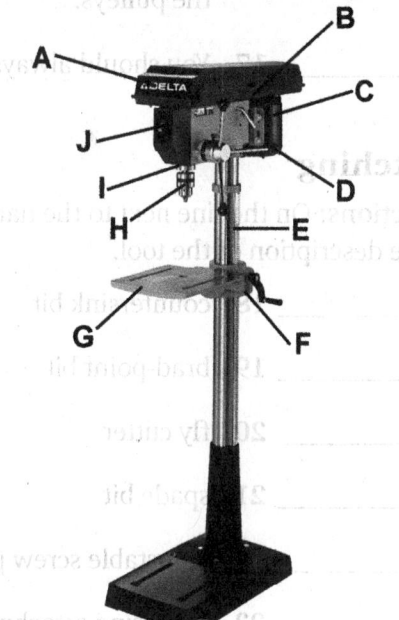

Fig. 27-1

Completion

Directions: On the line to the left of each sentence, write the word or phrase that correctly completes the sentence or answers the question.

_____ 11. A bit used to cut cross-grain and end-grain plugs and dowels is known as a(n) ___. (*two words*)

_____ 12. The ___ or stop controls how far the drill bit or cutting tool can move. (*two words*)

_____ 13. Use a(n) ___ to drill holes that are 1/4 inch or smaller. (*two words*)

_____ 14. To drill a hole in round (cylindrical) stock, you should first place the stock in a(n) ___.

(Continued on next page)

59

Name _____ Date _____ Class _____

True or False

Directions: Read each statement carefully. If the statement is true, write **True** in the blank to the left of that numbered item. If the statement is false, write **False** in the blank.

_____ 15. When used at the proper speed, a drill press can be used for sanding.

_____ 16. On machines with step pulleys, the speed is set by the way the belt is arranged on the pulleys.

_____ 17. You should always turn the power off before adjusting the depth stop.

Matching

Directions: On the line next to the name of each tool for the drill press, write the letter that corresponds to the description of the tool.

_____ 18. countersink bit

_____ 19. brad-point bit

_____ 20. fly cutter

_____ 21. spade bit

_____ 22. adjustable screw pilot bit

_____ 23. mortising attachment

_____ 24. multispur bit

_____ 25. hole saw

A. installs and countersinks flathead screws
B. fast-cutting bit that leaves a rather rough hole
C. cuts perfectly round, flat-bottomed holes
D. a bit with a sharp point that lets you place the hole exactly where you want it
E. cuts circular holes from 1 to 4 inches in diameter
F. cone-shaped bit that enlarges the top of a hole so that a flathead screw can be driven flush with the surface
G. cuts holes up to 6 inches in diameter
H. a drill bit surrounded by a four-sided chisel

60

Name _____ Date _____ Class _____

CHAPTER 29 — Sanders

Matching

Directions: Identify the parts of the belt-and-disc sander shown in Fig. 29-1. On the line next to the name of each part, write the letter from the illustration that shows the part.

_____ 1. tracking/tension lever

_____ 2. miter gauge

_____ 3. tilting table

_____ 4. belt sander

_____ 5. tilt gauge

_____ 6. disc sander

_____ 7. tilt release

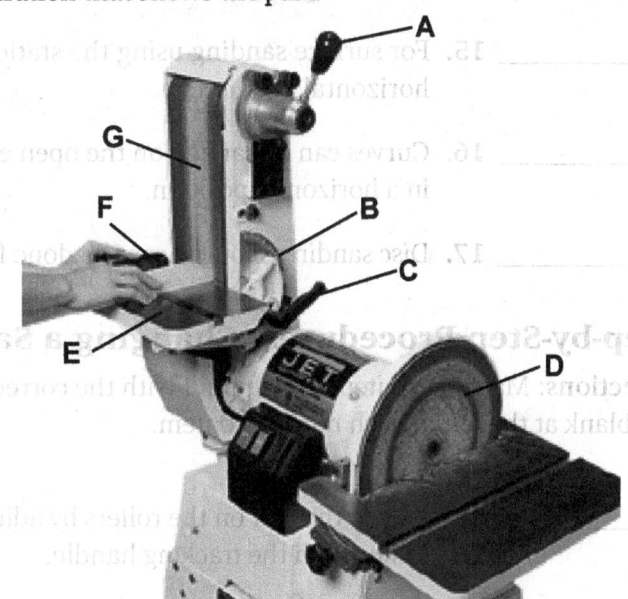

Fig. 29-1

Completion

Directions: On the line to the left of each sentence, write the word or phrase that correctly completes the sentence or answers the question.

_____ 8. A stationary ___ sander can be used in a vertical, horizontal, or slanted position.

_____ 9. A stationary ___ sander is used mostly for sanding end grain and the edges of a workpiece.

_____ 10. A sander that is good for sanding small parts and getting into hard-to-reach places is the ___ sander-grinder. (*two words*)

_____ 11. The oscillating ___ sander is a good choice for sanding curved or irregularly shaped edges.

_____ 12. When using a disc sander, you should hold the workpiece against the ___-moving side of the rotating disc.

(Continued on next page)

63

Name _____ Date _____ Class _____

True or False

Directions: Read each statement carefully. If the statement is true, write **True** in the blank to the left of that numbered item. If the statement is false, write **False** in the blank.

_____ 13. When using a power sander, you should adjust the table to 1/16 inch from the sanding belt or disc.

_____ 14. A good way to remove excess glue or finish from wood before refinishing is to sand it off using a belt-and-disc sander.

_____ 15. For surface sanding using the stationary belt sander, place the table in a horizontal position.

_____ 16. Curves can be sanded on the open end of a stationary belt sander when the table is in a horizontal position.

_____ 17. Disc sanding should never be done freehand.

Step-by-Step Procedures: Changing a Sanding Belt

Directions: Match each item in Column I with the correct step number in Column II. Write one letter in the blank at the left of each numbered item.

Column I

_____ 18. Center the belt on the rollers by adjusting the idler pulley with the tracking handle.

_____ 19. Remove the guards.

_____ 20. Remove the old belt and slip on a new one.

_____ 21. Increase the tension and replace the guards.

_____ 22. Apply a slight amount of tension.

_____ 23. Check the centering adjustment by moving the belt by hand.

_____ 24. Release the tension by turning the belt-tension knob.

_____ 25. Readjust the belt if necessary.

Column II
A. Step 1
B. Step 2
C. Step 3
D. Step 4
E. Step 5
F. Step 6
G. Step 7
H. Step 8

(Continued on next page)

64

Name _____ Date _____ Class _____

CHAPTER 30 — Lathe

Matching

Directions: Identify the parts of the lathe shown in Fig. 30-1. On the line next to the name of each part, write the letter from the illustration that shows the part.

_____ 1. lock for headstock position

_____ 2. tailstock spindle

_____ 3. spindle lock

_____ 4. headstock

_____ 5. faceplate

_____ 6. tailstock

_____ 7. tool rest

_____ 8. speed control

_____ 9. outboard bed

_____ 10. bed

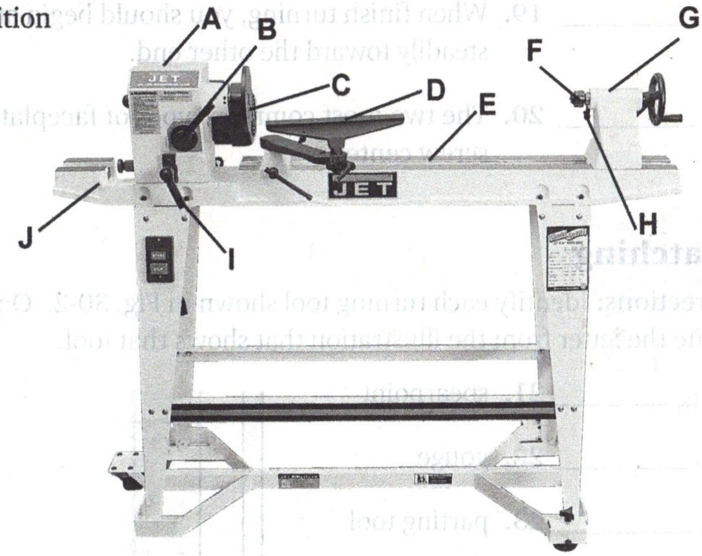

Fig. 30-1

Completion

Directions: On the line to the left of each sentence, write the word or phrase that correctly completes the sentence or answers the question.

_____ 11. In the ___ method of turning wood, the tool is held at right angles to the wood to remove fine particles of wood.

_____ 12. Thin shavings are produced as the tool peels away the waste stock in the ___ method of turning wood.

_____ 13. Another term for a live (moving) center is ___.

_____ 14. Turning wood between live and dead centers is known as ___ turning.

_____ 15. To make a bowl or other small circular object, you must mount a(n) ___ on the headstock spindle.

_____ 16. The appropriate tool for cutting a V is the small ___.

(Continued on next page)

65

Name _____ Date _____ Class _____

True or False

Directions: Read each statement carefully. If the statement is true, write **True** in the blank to the left of that numbered item. If the statement is false, write **False** in the blank.

_____ 17. Large-diameter stock should usually be turned at high speed.

_____ 18. Stock more than 3 inches square should be prepared for spindle turning by trimming off the edges to form an octagon shape.

_____ 19. When finish turning, you should begin at one end of the workpiece and move steadily toward the other end.

_____ 20. The two most common types of faceplates are the standard faceplate and the screw center.

Matching

Directions: Identify each turning tool shown in Fig. 30-2. On the line next to the name of each tool, write the letter from the illustration that shows that tool.

_____ 21. spearpoint

_____ 22. gouge

_____ 23. parting tool

_____ 24. skew

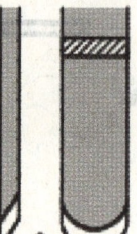

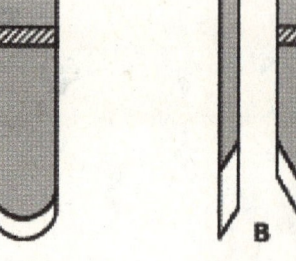

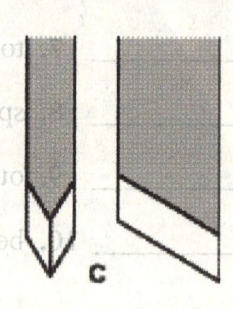

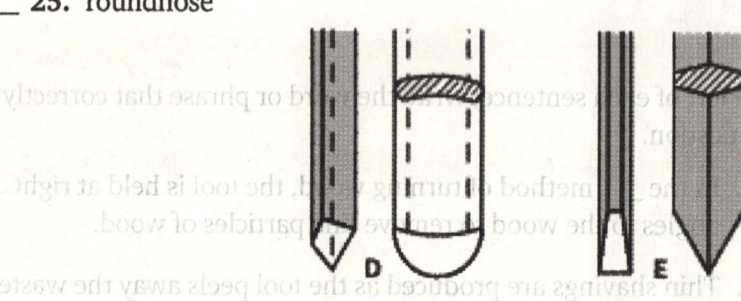

_____ 25. roundnose

Fig. 30-2

66

Name _____ Date _____ Class _____

CHAPTER 32
Applying Stains and Clear Finishes

Matching

Directions: On the line next to each type of finish, write the letter of the correct description.

_____ 1. linseed oil

_____ 2. tung oil

_____ 3. Danish oil

_____ 4. salad bowl finish

_____ 5. Deft® finish

A. commercial finish made of alkyd resins and tung oil that has been approved by the FDA as being completely foodsafe

B. semigloss, clear, interior wood finish made of tung oil and urethane

C. finish that can withstand hot dishes, but is not highly water-resistant and becomes sticky in damp weather

D. blend of oils and resins that penetrates, seals, and preserves wood finishes; requires only two coats

E. finish that penetrates deep into the wood and forms a long-lasting moisture barrier; requires several hand-rubbed coats

Completion

Directions: On the line to the left of each sentence, write the word or phrase that correctly completes the sentence or answers the question.

_____ 6. Stains created by adding color pigments to boiled linseed oil and turpentine are known as ___ stains.

_____ 7. ___ stains are made by mixing aniline dyes in oil. (*two words*)

_____ 8. The three types of varnish are alkyd resin, ___, and polyurethane.

_____ 9. Ash, hickory, and walnut are examples of ___-grained woods.

_____ 10. A(n) ___ made of one part shellac to seven parts denatured alcohol is used to seal wood that will be finished with shellac. (*two words*)

_____ 11. The process of changing the color of wood without changing its texture is known as ___.

_____ 12. Linseed oil and Danish oil are examples of ___ clear finishes.

_____ 13. To make your own liquid filler, you can use ___ to thin paste filler. (*two words*)

_____ 14. What type of stain will give the desired color and finish in one coat? (*two words*)

(*Continued on next page*)

69

True or False

Directions: Read each statement carefully. If the statement is true, write **True** in the blank to the left of that numbered item. If the statement is false, write **False** in the blank.

_____ 15. The end view shown in Fig. 32-1 shows wood to which a penetrating finish has been applied.

Fig. 32-1

_____ 16. To get a true idea of the final color of a project, you should apply the stain, sealer, filler, and topcoat to a piece of scrap wood or an inconspicuous part of the project.

_____ 17. Water-based stain can be made by mixing powdered aniline dye with hot water.

_____ 18. Shellac is a good finish for exterior projects because it is highly water-resistant.

_____ 19. To remove tiny air bubbles in varnish, allow the varnish to dry for several days before rubbing and polishing it with special abrasives.

Step-by-Step Procedures: Applying Lacquer with a Spray Gun

Directions: Match each item in Column I with the correct step number in Column II. Write one letter in the blank at the left of each numbered item.

Column I

_____ 20. Clean all equipment with lacquer thinner.

_____ 21. Make sure the spray gun is clean.

_____ 22. Spray on four or five coats of thin lacquer.

_____ 23. Clean the surface with a tack rag.

_____ 24. Put on your respirator mask, goggles, gloves, and clothing protection.

_____ 25. Test the spray gun on a piece of scrap stock.

Column II
A. Step 1
B. Step 2
C. Step 3
D. Step 4
E. Step 5
F. Step 6

(Continued on next page)

Name _____ Date _____ Class _____

CHAPTER 33
Applying Paints and Enamels

Completion

Directions: On the line to the left of each sentence, write the word or phrase that correctly completes the sentence or answers the question.

_____ 1. The amount of surface brightness a paint or enamel has is called ___.

_____ 2. ___ finishes are finishes that you cannot see through.

_____ 3. Bare wood should be prepared for painting by applying an undercoat of ___.

_____ 4. After applying oil-based paint or enamel, what solvent should you use to clean the equipment? (*two words*)

_____ 5. Which part of a table should you paint last?

_____ 6. Water-based ___ paint dries much more quickly than oil-based paint.

_____ 7. High-gloss paints that provide a slick, hard finish for easy cleaning are known as ___.

True or False

Directions: Read each statement carefully. If the statement is true, write **True** in the blank to the left of that numbered item. If the statement is false, write **False** in the blank.

_____ 8. Oil-based finishes usually don't have as strong an odor as water-based finishes.

_____ 9. Flat paint has no gloss.

_____ 10. Brushes used with latex paint should be cleaned using a solvent such as turpentine or mineral spirits.

_____ 11. When painting round or turned table legs, brush paint on using lengthwise strokes; do not brush all the way around them.

_____ 12. To finish a chair, you should turn the chair upside down and paint all the bottom surfaces first.

_____ 13. To finish a chest or cabinet, paint the panels on a chest or cabinet first, then the top, and then the moldings.

_____ 14. Paint is available as both brush-on and aerosol finishes, but enamel is sold only as a brush-on finish.

(Continued on next page)

71

Name _____ Date _____ Class _____

_____ 15. You should always paint in a well-ventilated area.

_____ 16. Drying times for paints and enamels vary from brand to brand, so it is important to read the label before using them.

_____ 17. It is usually not necessary to remove knobs, handles, and other hardware before painting furniture.

Step-by-Step Procedures: Applying Paint or Enamel

Directions: Match each item in Column I with the correct step number in Column II. Write one letter in the blank at the left of each numbered item.

Column I

_____ 18. Sand smooth with 320- or 400-grit sandpaper and clean the surface with a cloth to remove any sanding dust.

_____ 19. Prepare the surface by fixing dents or gouges, sealing knots, and sanding.

_____ 20. Stir the paint until smooth and well blended.

_____ 21. Apply an appropriate primer to any bare wood surfaces.

_____ 22. Apply a second coat in the same manner as the first, but do not sand it.

_____ 23. Make sure the surfaces to be painted are clean and free of dust particles.

_____ 24. Allow the surface to dry for at least 24 hours.

_____ 25. Using long, even strokes, apply the paint to the surface.

Column II
A. Step 1
B. Step 2
C. Step 3
D. Step 4
E. Step 5
F. Step 6
G. Step 7
H. Step 8

CHAPTER 34 Preparing for Construction

Matching

Directions: On the line next to each term, write the letter of the correct description.

_____ 1. abstract

_____ 2. building code

_____ 3. permit

_____ 4. survey

_____ 5. deed

_____ 6. certificate of occupancy

A. provides evidence of property ownership
B. traces ownership of property through legal documents
C. establishes minimum standards of quality and safety in housing construction
D. issued after a house has been inspected and is ready to be lived in
E. identifies boundaries of property
F. license that is required before construction can begin in areas that are covered by building codes

Completion

Directions: On the line to the left of each sentence, write the word or phrase that correctly completes the sentence or answers the question.

_____ 7. Fig. 34-1 shows an example of a(n) ___. (*two words*)

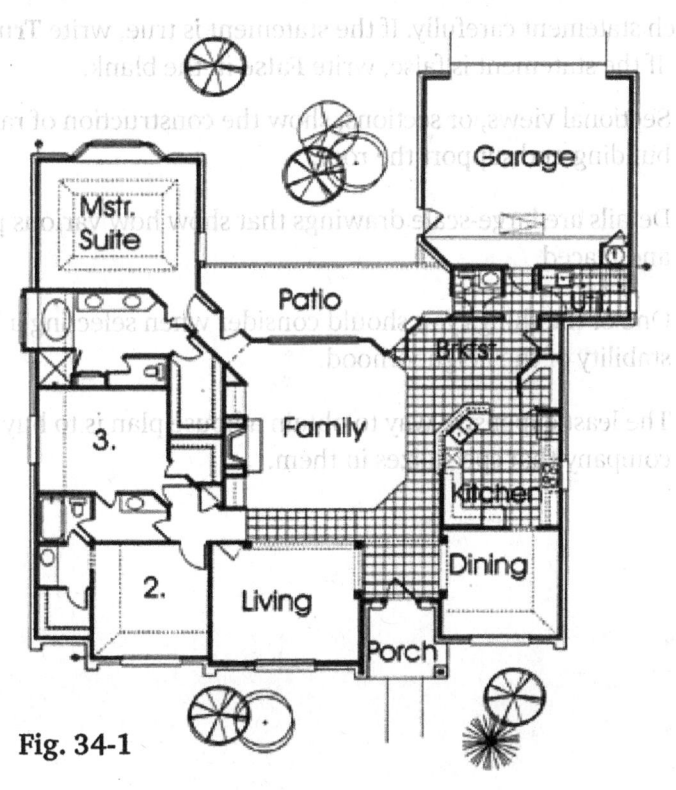

Fig. 34-1

(Continued on next page)

73

Name _____ Date _____ Class _____

_____ 8. All the materials needed to complete a construction project are named on the ___. (two words)

_____ 9. Another term for stick framing is ___ framing.

_____ 10. Housing that is built wholly or in part on an assembly line in a factory is known as ___ housing.

_____ 11. ___ parts are standard-sized parts that are made in a factory.

_____ 12. Written details on construction that are not described on floor plans or elsewhere are called ___.

_____ 13. Doors, windows, electrical circuits, and other items that would be hard to show on working drawings are represented by ___.

_____ 14. A top view of a building site that shows the precise location of footing and foundation walls is a(n) ___ plan.

_____ 15. A(n) ___ plan that shows a building site with boundaries, contours, existing roads, utilities, and other physical details.

_____ 16. A picture-like view of a building that shows exterior materials and the height of doors, windows, and rooms is called a(n) ___.

True or False

Directions: Read each statement carefully. If the statement is true, write **True** in the blank to the left of that numbered item. If the statement is false, write **False** in the blank.

_____ 17. Sectional views, or sections, show the construction of rafters used to span the building and support the roof.

_____ 18. Details are large-scale drawings that show how various parts are to be connected and placed.

_____ 19. One of the things you should consider when selecting a building lot is the future stability of the neighborhood.

_____ 20. The least expensive way to obtain a house plan is to buy a stock plan from a company that specializes in them.

CHAPTER 35 — Sitework and Foundations

Completion

Directions: On the line to the left of each sentence, write the word or phrase that correctly completes the sentence or answers the question.

_____ 1. What part of a house anchors it to the land and supports its weight?

_____ 2. A level like the one in Fig. 35-1, in which the telescope can be moved to different angles, is called a(n) ___ level.

_____ 3. The basic reference point for laying out a building using a level is called the ___. *(two words)*

_____ 4. A(n) ___ level has a telescope that is fixed in a horizontal position.

Fig. 35-1

_____ 5. To lay out a building on a site, the level is placed directly over the ___ mark, or point from which the layout will be sited.

_____ 6. Preparation of a building site, including surveying and clearing the land and laying out the building, is known as ___.

_____ 7. String stretched between ___ can be used to mark the outside line of foundation walls. *(two words)*

_____ 8. The foundation of a house rests on an enlarged base called a(n) ___.

_____ 9. The two types of foundation walls most commonly used for houses are poured concrete and ___. *(two words)*

_____ 10. Flat areas of poured concrete such as slab foundations and basement floors are known as concrete ___.

_____ 11. What connects foundation walls to the wood framing above them? *(two words)*

_____ 12. Concrete is made of fine aggregate, coarse aggregate, water, and ___.

_____ 13. Weak areas that form between concrete poured at different times are known as ___. *(two words)*

(Continued on next page)

75

Name _____ Date _____ Class _____

_____ 14. Fig. 35-2 shows a ___, which locates the future building accurately on the building site. *(two words)*

Fig. 35-2

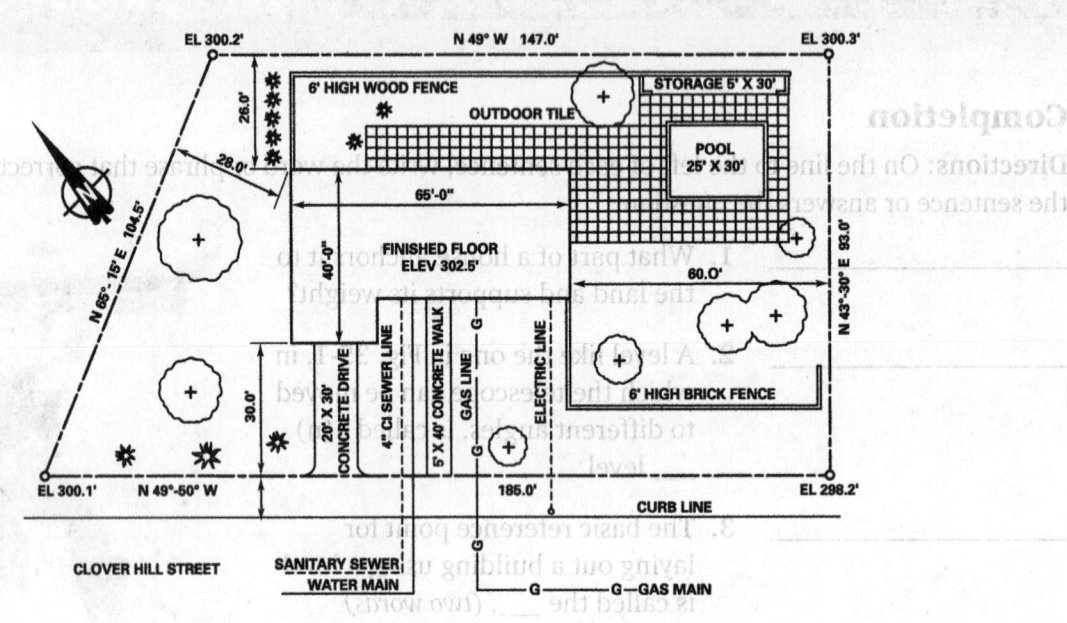

True or False

Directions: Read each statement carefully. If the statement is true, write **True** in the blank to the left of that numbered item. If the statement is false, write **False** in the blank.

_____ 15. The exact location of a building on a lot is affected by building codes, zoning codes, and natural features.

_____ 16. A foundation on a concrete slab does not need foundation walls.

_____ 17. The best way to pour a concrete foundation is to pour the lower half of the wall, wait 7 days for the lower half to cure, and then pour the upper half.

_____ 18. Unlike poured concrete foundations, concrete block foundations do not usually need formwork.

_____ 19. Mortar, or grout, is a mixture of coarse aggregate, fine aggregate, and water.

_____ 20. The finishing process for concrete depends on the intended use of the concrete.

Name _____ Date _____ Class _____

CHAPTER 36
Framing and Enclosing the Structure

Matching

Directions: Identify each type of roof shown in Fig. 36-1. On the line next to the name of each roof name, write the letter from the illustration that shows that type of roof.

_____ 1. gable roof

_____ 2. gable & valley roof

_____ 3. hip & valley roof

_____ 4. gambrel roof

_____ 5. flat roof

_____ 6. gable roof with dormer

_____ 7. hip roof

_____ 8. mansard roof

_____ 9. shed or lean-to roof

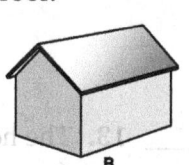

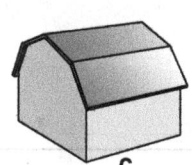

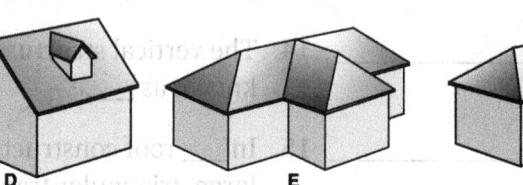

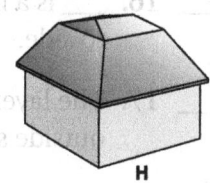

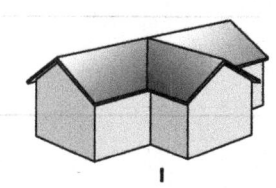

Fig. 36-1

Completion

Directions: On the line to the left of each sentence, write the word or phrase that correctly completes the sentence or answers the question.

_____ 10. The members in a conventionally framed house should be spaced *no more than* ___ inches apart.

_____ 11. What type of wood framing is shown in Fig. 36-2?

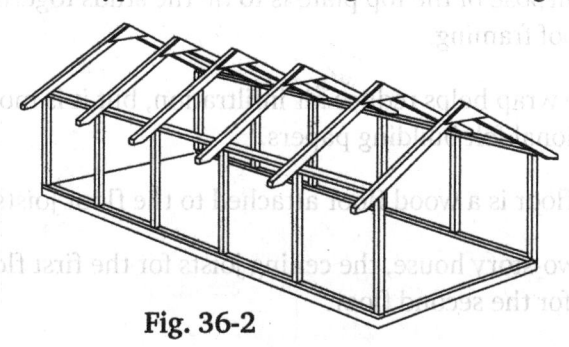

Fig. 36-2

(Continued on next page)

77

Name _____ Date _____ Class _____

_____ 12. What type of wood framing is shown in Fig. 36-3?

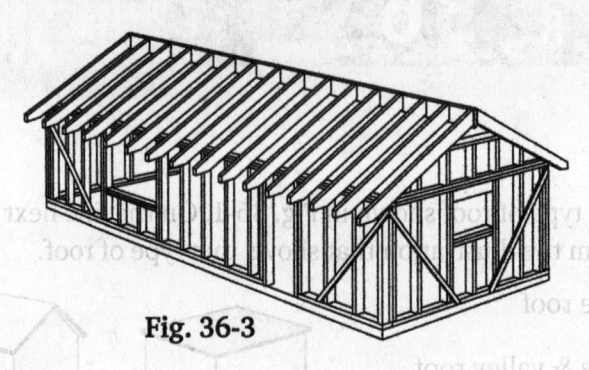

Fig. 36-3

_____ 13. The horizontal structural members that support a floor are the ___. (two words)

_____ 14. The vertical structural members that are placed between wall plates are known as ___.

_____ 15. In ___ roof construction, rafters and roof joists are prefabricated into large, triangular frames and delivered to the job site as a single unit.

_____ 16. ___ is a bracing method used to keep long joists from swaying from side to side.

_____ 17. The layer that is sometimes placed between a building's frame and the outside surface is called ___.

True or False

Directions: Read each statement carefully. If the statement is true, write **True** in the blank to the left of that numbered item. If the statement is false, write **False** in the blank.

_____ 18. Platform framing is a type of conventional framing.

_____ 19. The purpose of girders is to support trusses in trussed roof construction.

_____ 20. Interior walls may also be called *partitions*.

_____ 21. The purpose of the top plate is to tie the studs together and to connect the wall and roof framing.

_____ 22. House wrap helps reduce air infiltration, but it is more difficult to install than traditional felt building papers.

_____ 23. A subfloor is a wood floor attached to the floor joists under the finished floor.

_____ 24. In a two-story house, the ceiling joists for the first floor also serve as the floor joists for the second floor.

_____ 25. Inside walls are framed in the same way as outside walls.

Name _____ Date _____ Class _____

CHAPTER 37
Completing the Exterior

Matching

Directions: Identify each roofing element shown in Fig. 37-1. On the line next to the name of each item, write the letter from the illustration that shows that item.

_____ 1. starter strip

_____ 2. asphalt shingles

_____ 3. plywood sheathing

_____ 4. 15# asphalt saturated felt

_____ 5. metal drip strip

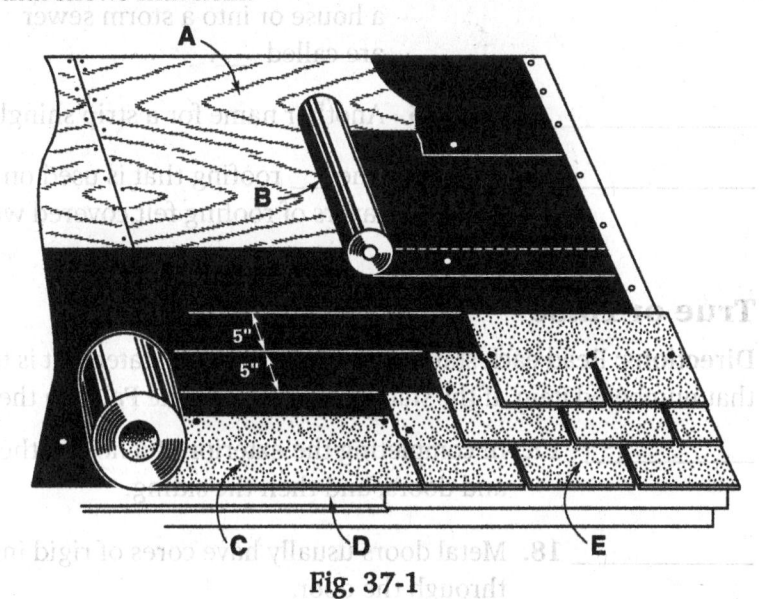

Fig. 37-1

Completion

Directions: On the line to the left of each sentence, write the word or phrase that correctly completes the sentence or answers the question.

_____ 6. The thin pieces of building material that are laid in overlapping rows on a roof are called ___.

_____ 7. The ___ is the assembly that supports a door and allows it to be connected to the house framing.

_____ 8. Roll roofing is an example of the ___ that is often installed before the exterior roofing materials are applied.

_____ 9. The troughs at the lower edge of a roof that catch water and carry it to downspouts are ___.

_____ 10. The purpose of louvered ___ is to provide ventilation for attic and roof spaces.

_____ 11. The part of a doorframe that fits inside the rough opening is the ___.

(Continued on next page)

79

Name _____ Date _____ Class _____

_____ 12. Figure 37-2 shows an example of ventilating ___.

_____ 13. A(n) ___ window is made of wood components that have been covered with rigid vinyl.

_____ 14. The pipes that carry water away from the foundation of a house or into a storm sewer are called ___.

Fig. 37-2

_____ 15. Another name for a strip shingle is ___ shingle.

_____ 16. The ___ roofing that is used on flat or low-pitched roofs consists of layers of roofing felt covered with asphalt and a top layer of gravel.

True or False

Directions: Read each statement carefully. If the statement is true, write **True** in the blank to the left of that numbered item. If the statement is false, write **False** in the blank.

_____ 17. In completing the exterior of a house, the roof is installed first, then the windows and doors, and then the siding.

_____ 18. Metal doors usually have cores of rigid insulation to help reduce heat loss through the door.

_____ 19. Steel is the most common structural element of a typical window.

_____ 20. The exterior wall covering of a house is called *siding*, regardless of whether the covering is wood, aluminum, vinyl, brick, or some other material.

80

CHAPTER 38 Completing the Interior

Matching

Directions: On the line next to each type of interior door, write the letter of the correct description.

_____ 1. bifold doors
_____ 2. sliding doors
_____ 3. pocket doors
_____ 4. louvered doors
_____ 5. folding doors
_____ 6. flush doors
_____ 7. panel doors

A. often used for closets because they provide ventilation
B. slide into a cavity in the wall to save space
C. made up of solid wood stiles, rails, and panels
D. hinged in the middle, these doors are often used to conceal closets, storage walls, and laundry areas
E. sometimes used as a temporary room divider
F. designed to cover unusually large openings efficiently
G. usually consist of a hollow core of light framework faced with thin plywood or hardboard

Completion

Directions: On the line to the left of each sentence, write the word or phrase that correctly completes the sentence or answers the question.

_____ 8. The piping used to bring water into a house from a public water system is the ___. *(two words)*

_____ 9. A heating and cooling system that is widely used in the southern regions of the United States is the ___. *(two words)*

_____ 10. Any material that slows down the transmission of heat or cold is known as ___.

_____ 11. ___ is an interior wall and ceiling finishing material made up of gypsum filler and faced with paper.

_____ 12. The vertical members that support the handrail and guardrail on a stairway are called ___.

_____ 13. At what stage is the piping run for plumbing in a new house?

_____ 14. The standard height for countertops is ___ inches.

_____ 15. One side of insulation is covered by a(n) ___ to prevent moisture from getting into interior walls, floors, or ceilings. *(two words)*

(Continued on next page)

Name _____ Date _____ Class _____

Matching

Directions: Identify each part of the stairway shown in Fig. 38-1. On the line next to the name of each part, write the letter from the illustration that shows that part.

_____ 16. nosing

_____ 17. stairwell header

_____ 18. finish floor line

_____ 19. tread

_____ 20. cutout stringer

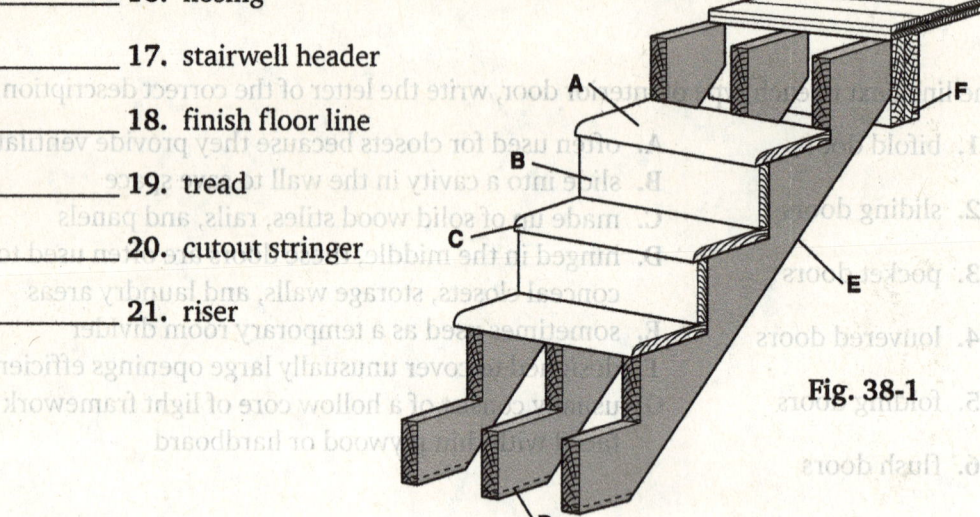

_____ 21. riser

Fig. 38-1

True or False

Directions: Read each statement carefully. If the statement is true, write **True** in the blank to the left of that numbered item. If the statement is false, write **False** in the blank.

_____ 22. The two stages of installing plumbing and electrical systems are the rough-in and close-in stages.

_____ 23. Unlike fresh-water pipes, soil pipes are under pressure to help remove wastes from the house.

_____ 24. A forced hot-air heating system consists of a furnace, ducts, and registers.

_____ 25. Drywall is available in special forms for water and fire resistance.

(Continued on next page)

82

Name _____ Date _____ Class _____

CHAPTER 39
Remodeling and Renovation

Completion

Directions: On the line to the left of each sentence, write the word or phrase that correctly completes the sentence or answers the question.

_____ 1. Changing a house to meet the needs of new owners or make it more suitable for them is known as ___.

_____ 2. Before undertaking extensive remodeling, homeowners should check to make sure that the proposed changes are ___ (worth the money required to make them). *(two words)*

_____ 3. The slender wood strips that were used as a base for plaster in older houses are known as ___. *(two words)*

_____ 4. Before purchasing an older home, you should have a professional ___ examine the house and point out any problems that may need to be addressed after the purchase. *(two words)*

_____ 5. When you add a room to a house, you should design the roof over the new room to duplicate the existing roof or to ___ with it.

_____ 6. Before building, be sure to check ___ and zoning laws to make sure your plans will meet all requirements. *(two words)*

_____ 7. Outside walls of a house are ___ walls that help support the weight of the structure.

_____ 8. Many homeowners choose to hide wall imperfections, especially in basements, using wood or plywood ___.

_____ 9. A(n) ___ ceiling consists of a grid system that hangs from an existing roof using a series of runners, cross-tees, and molding.

_____ 10. When a floor is made of concrete, a(n) ___ must be installed between the concrete and hardwood finish flooring. *(two words)*

_____ 11. ___ involves restoring an older house to its original condition, often because of its historical or architectural importance.

_____ 12. The removal of old building materials so that rebuilding can begin is known as ___.

_____ 13. Before a load-bearing wall can be removed, it must be ___, or temporarily supported. *(two words)*

(Continued on next page)

Name _____ Date _____ Class _____

True or False

Directions: Read each statement carefully. If the statement is true, write **True** in the blank to the left of that numbered item. If the statement is false, write **False** in the blank.

_____ 14. Some remodeling projects concentrate on replacing older windows with new ones to increase the energy efficiency of a house.

_____ 15. Remodeling can change the number of rooms in a house.

_____ 16. It takes more time and money to repair drywall than to repair plaster.

_____ 17. Before carpeting or vinyl flooring can be used in a basement, you should check to be sure that it can be used below grade.

_____ 18. Building codes specify how close an addition can be to the property line.

_____ 19. For many families, a design in which a bedroom opens onto a dining area is a good idea.

_____ 20. Whenever possible, you should start a door opening at a stud so that you will not have to remove more than two studs to install the door.

Name _____ Date _____ Class _____

CHAPTER 40
Maintenance and Repair

Completion

Directions: On the line to the left of each sentence, write the word or phrase that correctly completes the sentence or answers the question.

_____ 1. A material that is similar to caulk but is more flexible (and more expensive) is ___.

_____ 2. When inspecting exterior walls, you should make sure that soil comes no closer than ___ inches to any wood.

_____ 3. The problem shown in Fig. 40-1 is known as ___. *(two words)*

Fig. 40-1

_____ 4. When exterior siding alternately swells and shrinks as the moisture behind it is absorbed and then evaporates, a paint problem known as ___ results.

_____ 5. To correct a mildew problem, scrub the surface with a solution of ⅓-cup of ___ and ½ cup of household bleach mixed with 1 gallon of warm water. *(two words)*

_____ 6. Heat loss by leakage of warm air through cracks in windows, walls, and doors is known as ___.

_____ 7. Joints between two dissimilar materials on the exterior of a house should be sealed with ___ or a sealant.

_____ 8. The process of repairing the mortar in a brick wall is known as ___.

_____ 9. The ___ of a fireplace allows the flue to be closed when the fireplace is not in use.

_____ 10. A rust-resistant screen known as a(n) ___ may be required for fireplaces on or near combustible roofs, woodland, lumber, or other combustible materials. *(two words)*

(Continued on next page)

85

Name _____ Date _____ Class _____

_____ 11. If you found a tiny pile of fine, flour-like powder near a wood doorframe, which insect would you suspect has infested and damaged the doorframe? *(two words)*

_____ 12. The two most common types of termites in the United States are the ___ and subterranean termites.

True or False

Directions: Read each statement carefully. If the statement is true, write **True** in the blank to the left of that numbered item. If the statement is false, write **False** in the blank.

_____ 13. Chimney cleaning can easily be done by homeowners.

_____ 14. Mildew thrives in high temperature and low humidity.

_____ 15. On cracks 1/16 inch wide or larger, you should undercut the crack to an inverted V-shape before patching.

_____ 16. You should check a basement for damp spots at least once each season.

_____ 17. Eye and skin protection is always necessary when working with chemicals such as trisodium phosphate.

_____ 18. Tooling a caulk or sealant is the process of pushing it into the joint and smoothing it over.

_____ 19. Leakage of cold outside air into a house is known as *exfiltration*.

Step-by-Step Procedures: Repairing Localized Peeling

Directions: Match each item in Column I with the correct step number in Column II. Write one letter in the blank at the left of each numbered item.

Column I	Column II
_____ 20. Scrape off the old paint from the affected area and for about 12 inches around the area.	A. Step 1
	B. Step 2
_____ 21. Apply a top coat of recommended house paint according to label directions.	C. Step 3
	D. Step 4
	E. Step 5
_____ 22. Spot-prime with a recommended undercoat.	F. Step 6
_____ 23. Sand the surface to fresh wood.	
_____ 24. Seal all seams, holes, and cracks against moisture with caulk.	
_____ 25. Locate and eliminate all sources of moisture.	

86

Name _____ Date _____ Class _____

Advanced Woodworking Techniques

Completion

Directions: On the line to the left of each sentence, write the word or phrase that correctly completes the sentence or answers the question.

_____ 1. Excess glue that escapes a joint while it is clamped is called adhesive ___.

_____ 2. Clamping ___ are used to distribute pressure across the width and length of the stock.

_____ 3. A workpiece that includes several unusual angles can be held together using a(n) ___ clamp.

_____ 4. A(n) ___ cut does not go all the way through the workpiece from one end to the other.

_____ 5. A(n) ___ pin provides leverage the operator can use to guide the workpiece as it is routed.

_____ 6. ___ and ___ form the perimeter of a face frame.

_____ 7. A table saw ___ fully supports the workpiece and is easier to use for large pieces of stock.

_____ 8. A ___ miter is cut across the width of the wood.

_____ 9. Another term for slip joint is ___ joint.

_____ 10. A sliding dovetail joint is made with a pin that fits into a(n) ___.

_____ 11. A(n) ___ is used to trim the parts of a pen to length.

_____ 12. When turning a pen, set the lathe to about ___ rpm.

_____ 13. To create a durable finish on a pen, use ___ polish.

(Continued on next page)

Name _____ Date _____ Class _____

True or False

Directions: Read each statement carefully. If the statement is true, write **True** in the blank to the left of that numbered item. If the statement is false, write **False** in the blank.

_____ 14. The best way to accumulate a set of clamps is to purchase the clamps as they are needed.

_____ 15. Right-angle clamps hold items together at a 90-degree angle.

_____ 16. *Never* position the workpiece on a router table so the cutting action is at the outer edge of the stock.

_____ 17. Wood that has a slight crown cannot be routed using a router table.

_____ 18. During routing, a pattern must have smooth edges; otherwise, irregularities will be duplicated in the workpiece.

_____ 19. Mortise-and-tenon joints are the joints most commonly used for face frames.

_____ 20. If ¾-inch plywood is being used for a cabinet carcase, a rabbet cut in it should be no more than ¼-inch deep.

_____ 21. Splines increase the amount of face-grain glue surface.

_____ 22. A slip joint requires a mortise.

_____ 23. A rabbet-and-dado joint is stronger for joining the front of a drawer to the sides than a sliding dovetail joint.

_____ 24. When assembling a pen, first lay the components out in the correct order.

_____ 25. No attempt should be made to match the grain of pen components.

(Continued on next page)

Name _____ Date _____ Class _____

Science Activity 1

Water Evaporation from Wood

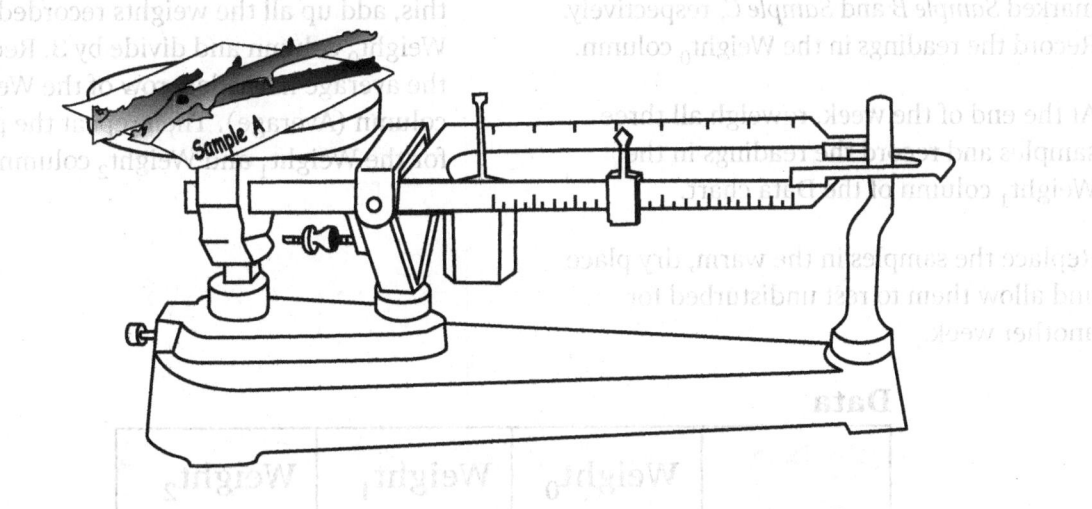

A living tree has a large amount of moisture in its fibers. After the tree is cut, water begins to evaporate from it as the wood dries out. As the water evaporates, the dimensions of the wood change slightly. This is why wood should be dried or seasoned before it is used to make furniture and other products.

How rapidly does the water evaporate from a freshly cut tree? You can find out in this investigation.

Procedure

1. Cut three pieces of nonporous paper to fit on the pan of the laboratory balance. Label them *Sample A, Sample B,* and *Sample C.*

2. Break or cut up one of the tree branches, being sure you have a mass of at least 500 grams after the leaves are removed. Place the pieces on the paper marked *Sample A.*

3. Weigh the pieces on the laboratory balance. Record the reading as accurately as possible in the appropriate place in the Weight$_0$ column of the Data chart on page 90.

MATERIALS

- three large, freshly cut branches weighing at least 500 grams each
- laboratory balance (can be obtained from the science department)
- nonporous paper (if paper designed for use on a balance is not available, waxed paper can be used)

(Continued on next page)

Name _____ Date _____ Class _____

Science Activity 1 (continued)

4. Leaving the wood on the paper, place *Sample A* in a warm, dry location and allow it to rest undisturbed for one week.

5. Repeat Steps 2-4 for the other two tree branches, but place them on the papers marked *Sample B* and *Sample C*, respectively. Record the readings in the Weight$_0$ column.

6. At the end of the week, reweigh all three samples and record the readings in the Weight$_1$ column of the Data chart.

7. Replace the samples in the warm, dry place and allow them to rest undisturbed for another week.

8. Weigh all three samples again and record the readings in the Weight$_2$ column of the Data chart.

9. Compute the average of the weights of the three samples listed in each column. To do this, add up all the weights recorded in the Weight$_0$ column and divide by 3. Record the average in the last row of the Weight$_0$ column (Average). Then repeat the process for the Weight$_1$ and Weight$_2$ columns.

Data

	Weight$_0$	Weight$_1$	Weight$_2$
Sample A			
Sample B			
Sample C			
Average			

Conclusions

1. What happened to the average weight of the samples over the period of time you tested? Why? _____

2. Why was it important to test three samples instead of only one? _____
_____

3. What can you conclude about the water content of the branches after a period of two weeks?
_____

4. How could you change this experiment to test water evaporation from wood more thoroughly?
_____
_____

Name _____ Date _____ Class _____

Science Activity 2

Testing Wood Preservatives

In recent years, scientists have become increasingly concerned about the safety of some wood preservatives. A *preservative* is a chemical applied to wood to help prevent it from rotting or decaying. The EPA (Environmental Protection Agency) has been reviewing major commercial preservatives such as creosote and arsenics. They are trying to determine if these chemicals are serious health hazards.

It has been proven experimentally that certain preservatives are carcinogenic (cancer-causing). The greatest possibility for health damage exists for those workers who apply the preservatives. However, some of these chemicals can be absorbed through the skin when the wood is touched. To prevent this, the EPA suggests sealing any treated objects—such as outdoor furniture—that might frequently be in contact with skin. Varnish or shellac can be used for this purpose.

What common household chemicals make good wood preservatives? In this investigation, you will test the preservative qualities of three common chemicals: sodium chloride (table salt), acetone (nail polish remover), and sodium hypochlorite (household bleach).

MATERIALS

- sodium chloride solution (prepare by dissolving 2-tablespoons of table salt in 1 cup of warm water)
- acetone (nail polish remover)
- sodium hypochlorite (household bleach)
- 5 pieces of wood, each approximately 4" × 4" × 4"
- 5 jars or beakers, each large enough to hold one of the wood pieces
- tongs
- gloves

Procedure

1. Label 5 jars or beakers as follows:
   - A – Sodium Chloride
   - B – Acetone
   - C – Sodium Hypochlorite
   - D – Water
   - E – No Treatment

2. Place one piece of wood in each beaker.

3. Fill beaker A with enough sodium chloride solution to completely cover the wood.

4. Follow the same procedure with beakers B, C, and D, filling the beakers with the solution named on the label. **NOTE:** Acetone and sodium hypochlorite are easily absorbed through the skin and may cause burns or other health problems. Do not allow these chemicals to touch your skin. Wear appropriate gloves while handling them.

5. Leave beaker E (No Treatment) without any liquid.

6. Place all five beakers in a warm, dry place away from light and drafts. Allow them to remain overnight.

7. After the soaking time, line up the beakers so that you can see the labels. Place several layers of absorbent paper toweling in front of each beaker.

(Continued on next page)

91

Name _____ Date _____ Class _____

Science Activity 2 *(continued)*

8. Using the tongs, remove each piece of wood from its beaker and place it on the absorbent toweling. You may want to place a second label in front of each beaker so that you'll be able to tell which piece of wood came from which beaker.

9. Dispose of the chemicals in the beakers, but do not remove the labels.

10. Using the tongs, place each piece of wood back into the same beaker it was in originally, being careful not to mix up the samples.

11. Place the beakers in a warm, damp place—a dark closet, for example—for two or three weeks, if possible.

12. At the end of the two or three weeks, observe each piece of wood carefully. Note any mold (such as might be found on a piece of bread in a refrigerator) or other signs of decay. Record the appearance of each piece in the Data chart below.

NOTE: Put on appropriate gloves before handling the treated wood. Do not handle the wood with your bare hands. You may choose to use tweezers or dissecting picks, if available, to probe the wood gently.

Data

Beaker	Description of Wood
A – Sodium Chloride	
B – Acetone	
C – Sodium Hypochlorite	
D – Water	
E – No Treatment	

Conclusions

1. Which of the chemicals tested would be the best preservative for wood?

_____

2. Explain the purpose of testing one piece of wood in plain water and one without any liquid.

_____
_____
_____

3. Based on your results from this investigation, what other household chemicals might you test as wood preservatives? Why?

_____
_____
_____

(continued on next page)

92

Name _____ Date _____ Class _____

Science Activity 3

Determining Wood Density

Density is the property of a material that makes it feel heavy or light for its size when you pick it up. The heavier it feels, the denser it is. Imagine blocks of wood, steel, and Styrofoam all the same size. Which would feel heaviest? Lightest? Steel is denser than wood, which is denser than Styrofoam.

Materials with lower density also may be softer and more easily deformed by scratching or cutting. You might be able to scratch a piece of wood, and you could definitely scratch Styrofoam, but not a block of steel.

You can easily guess how the densities of some kinds of wood, such as oak and balsa, compare. What about pine, maple, and spruce? To find out exactly, you'd have to measure the density of each.

In this investigation, you will determine the density of samples of several kinds of woods. *Density* is a scientific term, defined as the mass per unit of volume of a substance. Scientists usually express density in metric terms as *grams per cubic centimeter*. To find the density of a substance, first you determine its mass and volume. Then divide the mass by the volume. The result is the density. The higher the number, the more dense the substance is.

MATERIALS
- blocks of spruce, poplar, oak, pine, and other wood in various sizes
- metric ruler
- metric scale

Procedure

1. See if you can predict, by heft alone, the relative density of each kind of wood. List your predictions below with the most dense wood at the top and the least dense at the bottom. Then carry out Steps 2–5 for each block of wood, recording your measurements and the results of your calculations in the Data chart on page 94.

    Predictions:

2. Measure each block's thickness, length, and width to the nearest whole millimeter using the metric ruler. Record these measurements in the Data chart.

3. Multiply the three measurements (thickness × length × width) to find the volume. Record the volume.

4. Find the mass of the block in grams. Record the mass.

5. Divide the mass by the volume. This will give you a very close approximation of the wood's density.

(Continued on next page)

Name _____ Date _____ Class _____

Science Activity 3 (continued)

Data

Kind of Wood	Thickness	Width	Length	Volume	Mass	Density

Conclusions

1. Which of the woods you tested is the most dense? _____

2. Which kind is the least dense? _____

3. According to your results, rank the wood samples from highest to lowest density.

_____

_____

4. Compare your results with your predictions. How close were your predictions? _____

_____

_____

94

Name _____ Date _____ Class _____

Science Activity 4

Viscosity of Wood Adhesives

Cohesion occurs when *like* molecules bond together. The molecules in a drop of epoxy glue, for example, are held together by cohesion. *Adhesion* takes place when *unlike* molecules bond together. An example is when epoxy glue adheres to a piece of pine. The wood has one type of molecule and the glue has another type. As the epoxy cures, both its cohesive and adhesive qualities grow stronger.

Some adhesives are thick and pour very slowly. The rate at which a liquid pours is a measure of its *viscosity*. High viscosity means a high level of cohesion between the molecules in the liquid.

Your textbook discusses the adhesive strength of several wood glues. For this activity you will measure their cohesive strength by observing their viscosity.

MATERIALS

- 4 small, clear plastic cups
- 4 different types of wood adhesives
- stopwatch
- waxed paper

Procedure

1. Label 4 small, clear plastic cups with the names of the 4 glues you will be testing.

2. Place the cups on 4 pieces of waxed paper large enough to cover the mouth of the plastic cup when the cup is turned over.

3. In the bottom of the first cup, place an amount about 1" (2.5 cm) in diameter of the type of adhesive written on its label. Place the adhesive as close to the center of the bottom of the cup as possible.

4. Have the stopwatch ready. Ask a helper to turn the cup over so that it is face-down on the paper. As the cup touches the paper, begin timing.

5. Run the stopwatch until the adhesive in the cup has either dripped or run down the side of the cup to touch the paper.

6. In the Data chart on page 96, record the time it took the adhesive to reach the paper.

7. Repeat Steps 2 to 6 three more times, once for each of the remaining adhesives.

8. In the last column of the Data chart, rank the adhesives from highest viscosity to lowest, based on your experimental results. Assign a value of 1 to the adhesive with the highest viscosity and a value of 4 to the one with the lowest viscosity.

(Continued on next page)

Name _____ Date _____ Class _____

Science Activity 4 (continued)

Data

Name of Adhesive	Time	Relative Viscosity

Conclusions

1. Which of the adhesives you tested is the most viscous? _____

2. Which kind is the least viscous? _____

3. Think about your results and the discussion of adhesion and cohesion on page 95. Which of the following do you think would create the strongest bond between two pieces of wood: a thick layer of a very viscous adhesive, or a thin layer of a less viscous adhesive? Why?

_____
_____
_____
_____
_____
_____

96

Name _____ Date _____ Class _____

Science Activity 5

Muscle Force for Planing

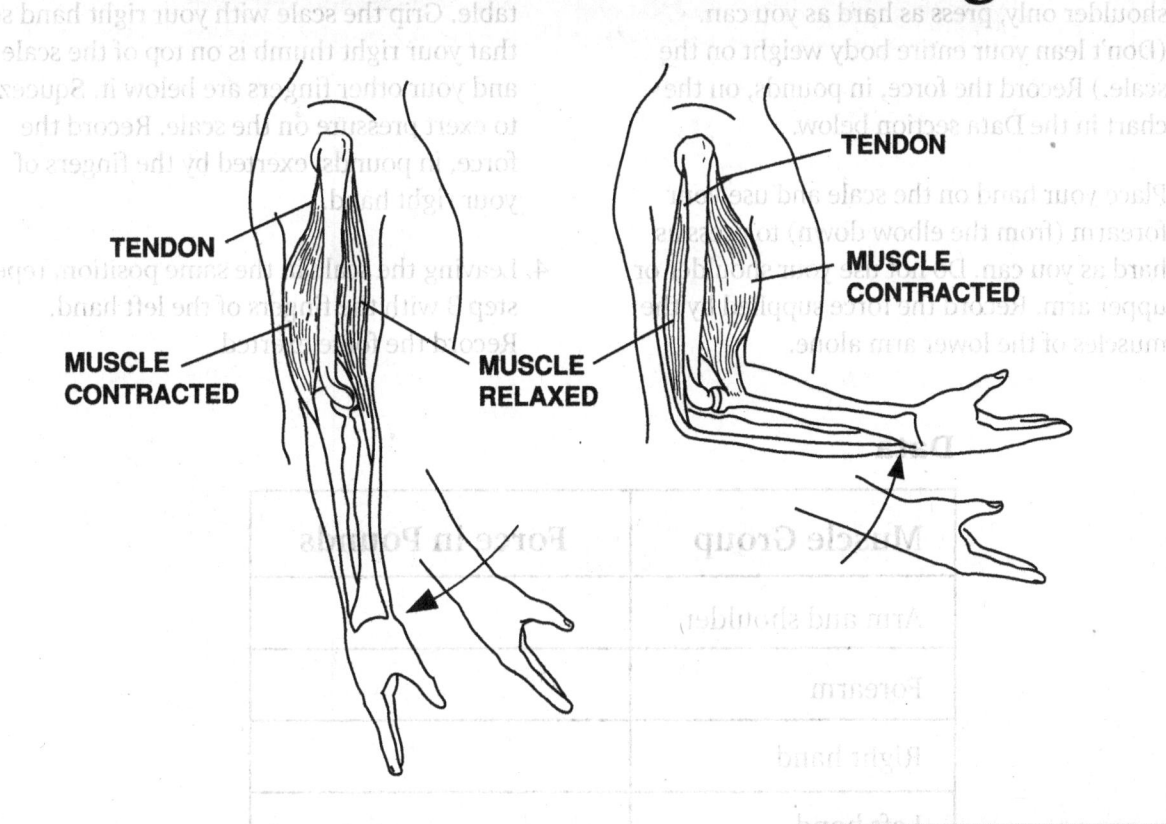

TENDON
MUSCLE CONTRACTED
MUSCLE RELAXED
TENDON
MUSCLE CONTRACTED

When you plane a piece of wood, the motion of your arm is directed by signals from your brain. These signals travel down through the spinal cord through the nerves of your arm to the muscles you are using. The muscles are made of bundles of hair-thin muscle fibers that contract. These fiber bundles are connected to the joints by tough bands of tissue called *tendons,* as shown in the drawing.

Your muscles contract and relax in groups to allow you to move a plane or other tool back and forth. To provide a lot of strength, many fibers contract. For less power, fewer fibers contract. Strong muscles have many more fibers and blood vessels than weak muscles. The more you use the muscles, the more fibers your muscles produce.

In hand woodwork, the muscles of the hand and arm must supply the movement and force necessary to shape, cut, and drill wood. How strong are these muscles? You can find out in this simple investigation.

MATERIALS
- bathroom scale
- table

(Continued on next page)

97

Name _____ Date _____ Class _____

Science Activity 5 (continued)

Procedure

1. Lay the bathroom scale on a table. Place your hand on the scale. Using your arm and shoulder only, press as hard as you can. (Don't lean your entire body weight on the scale.) Record the force, in pounds, on the chart in the Data section below.

2. Place your hand on the scale and use your forearm (from the elbow down) to press as hard as you can. Do not use your shoulder or upper arm. Record the force supplied by the muscles of the lower arm alone.

3. Slide the scale toward you so that 3-4 inches of the scale extend beyond the edge of the table. Grip the scale with your right hand so that your right thumb is on top of the scale and your other fingers are below it. Squeeze to exert pressure on the scale. Record the force, in pounds, exerted by the fingers of your right hand.

4. Leaving the scale in the same position, repeat step 3 with the fingers of the left hand. Record the force exerted.

Data

Muscle Group	Force in Pounds
Arm and shoulder	
Forearm	
Right hand	
Left hand	

Conclusions

1. Which step produced the most force?
   Why? _____

2. The fingers of which hand, left or right, exerted the most force?
   Why? _____

3. What is the difference between the way you use your arm muscles for planing and for sawing? _____
   _____

4. In which task do more muscle fibers contract? _____
   _____
   _____

(Continued on next page)

98

Name _____ Date _____ Class _____

Science Activity 6

Effect of Acid Rain on Plants

One of the current concerns of scientists is to understand more fully the impact of air pollution on the growth of our forests. The burning of fossil fuels (gasoline, natural gas, and coal) damages living plants. Botanists (scientists who study plants) have found forests in Colorado, Vermont, the Carolinas, and Georgia that have been damaged by acid rain. *Acid rain* is a chemical formed when gases such as sulfur dioxide or nitrogen monoxide combine with rain. Softwoods have been most affected.

Chemists have been able to create a mixture of chemicals identical to those found in acid rain. They have applied these chemicals to young seedlings and recorded the results. The experimental results prove that acid rain is a probable cause of the early death of many trees.

The *pH* of a liquid is a measure of its acidity. The pH scale runs from 1 to 14. In general, liquids with a pH of less than 7 are considered *acids,* and those with a pH of more than 7 are *bases.* Liquids with a pH of 7 are said to be *neutral.*

In this investigation, you will determine the acidity of rainwater and groundwater in your area. You will also test the effect of the acidity of water on young seedlings or plants.

MATERIALS
- 3 empty, clean jars
- pH paper (or red and blue litmus paper)
- medicine dropper
- vinegar
- 1 aspirin tablet
- soapy water
- tap water
- clean containers
- 3 small, healthy seedlings (young house plants can be substituted if seedlings are not available)

Procedure

1. During a rain shower, collect a sample of rain in one of the jars. If possible, collect rainwater samples from more than one location. Collect a sample of water from a pond, lake, or stream in each of the other jars.

2. One reliable way to test the pH of a liquid is to use pH paper. For practice, using a medicine dropper, place a small amount of vinegar on a piece of pH paper. Check the color of the paper against the scale on the pH paper package. Record the pH in the Data chart on page 100.

3. Wash the medicine dropper thoroughly to remove the vinegar. Then repeat the procedure in Step 2, but use an aspirin dissolved in 2 tablespoons of water instead of vinegar. Record its pH.

4. Repeat Step 3, washing the dropper carefully, and test a small amount of soapy water with the pH paper. Record its pH.

5. Now test each of the water samples you collected in Step 1. Record the pH of each sample in the Data chart.

(Continued on next page)

Name _____ Date _____ Class _____

Science Activity 6 (continued)

6. Label the three seedlings as follows:
   - Acidic Water
   - Basic Water
   - Tap Water

7. For a period of 3 weeks, water the seedlings on a regular basis using the following solutions. (Be sure to mix well before using.)
   - For the seedling labeled *Acidic Water,* use a solution made of 1 teaspoon white vinegar to 1 cup water.
   - For the seedling labeled *Basic Water,* use a solution of 1 teaspoon mild dish soap to 1 cup water.
   - For the seedling labeled *Tap Water,* use water straight from the faucet.

8. After 3 weeks, test the pH of the soil for each seedling. Record the pH in the Data chart.

9. Observe each of the seedlings carefully. Record your observations in the Data chart.

Data

Tests	pH	Observations
pH Paper Tests		
Vinegar		
Aspirin		
Soapy Water		
Plants		
Acidic Water		
Basic Water		
Tap Water		

Conclusions

1. How did the pH of water affect the seedlings? _____

_____

2. Based on your observations, what effect might acid rain have on seedlings and young trees?

Why? _____

_____

100

Name _____ Date _____ Class _____

Science Activity 7

Analyzing Tree Rings

For many years scientists have known that a single tree ring (also called an *annual growth ring*) indicates approximately one year's growth of a tree. Each new ring forms just below the bark, so the outer rings are the most recent.

The size and other characteristics of the rings offer accurate information not only about the age of the tree, but also about climatic conditions during the life of the tree. In fact, tree rings are a very reliable source of climatic information because they reflect a large number of variables (drought, pest infestations, etc.) over a large part of the planet.

For trees that have been cut down, it is easy to observe the annual growth rings. It would not be worthwhile to cut down a large number of trees just to study the climate, however. Fortunately, it is possible to get the same information without killing the tree. Scientists have developed a system in which they bore into a tree using a hollow pipe. The pipe is inserted through the center of the tree, usually in two or three different places near the tree's base. This procedure does not harm the living tree. The resulting sample is called a *core*. A core taken through the center of a tree shows a cross-section that includes parts of all the annual growth rings.

The most obvious information gained from coring a tree is the age of the tree and the climatic conditions during the tree's life to date. Because a ring forms about once a year, the number of rings approximates the age of the tree. The width of the rings tells about the climate. Wide rings are made during years of favorable growth conditions—the right amount of rainfall, absence of pests and diseases, and proper temperatures for the species of tree. Narrow rings occur during drought or other unfavorable conditions.

MATERIALS

- cores or sections from tree trunks ranging from 4" (10 cm) to 12" (30 cm) in diameter and from 2 or 3 different types of trees.
- handheld magnifying glass
- ruler

Procedure

1. Examine each of the tree samples closely, using both your unaided eye and the magnifying glass, paying special attention to the number and spacing of growth rings.

2. Complete the first 5 columns of the Data table for each sample, using the ruler as required. Use a separate sheet of paper to continue the chart if necessary.

(Continued on next page)

101

Name _____ Date _____ Class _____

Science Activity 7 *(continued)*

3. In the last column of the Data chart, record your estimation of the climatic conditions over the lifetime of each sample. Include relative rainfall amounts. (Which years had the most rain?)

4. Check with your local library or on the Internet to find the actual rainfall during the corresponding years. Record the information for each sample on a separate sheet of paper and attach it to this sheet.

Data

Sample No.	Type of Tree	Diameter	No. of Rings	Approx. Age	Climatic Summary

Conclusions

1. Consider your findings and the information you recorded about your local rainfall amounts. On average, do the annual growth rings indicate the rainfall correctly? Explain.

_____

_____

_____

2. Compare the growth rings for the same year in samples from different types of trees. What information does this give you about the rainfall needed by these types of trees?

_____

_____

102

Name _____ Date _____ Class _____

Science Activity 8

Leaf Structure and Tree Growth

LEAF STOMATA

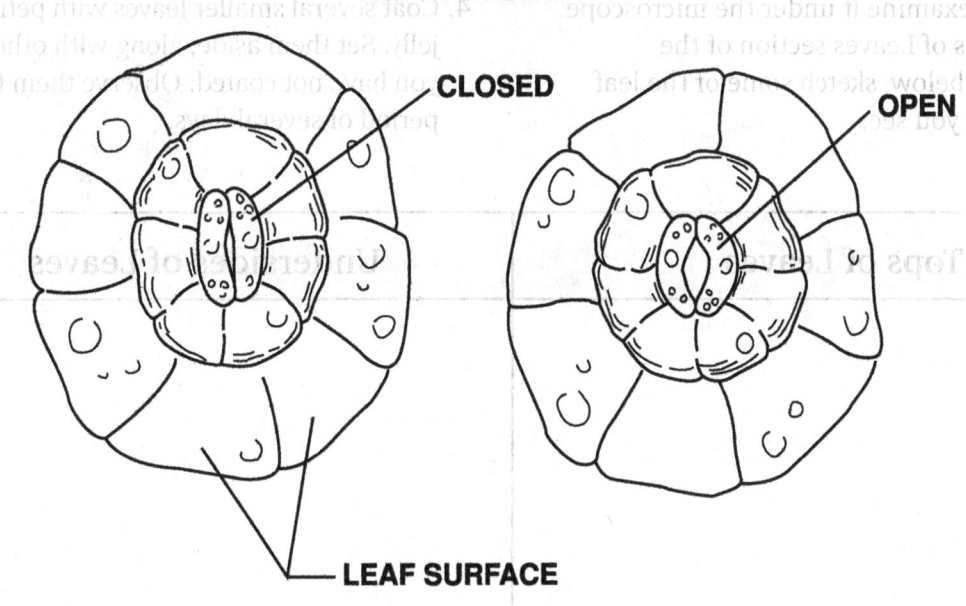

CLOSED   OPEN

LEAF SURFACE

The green leaves of a tree carry out photosynthesis to make sugar, which they use as food. To do this, they use carbon dioxide from the air, water from the soil, and the energy of sunlight. How does the leaf get carbon dioxide? Tiny oval openings in the leaf's surface called *stomata* let in air. The cells inside the leaf come into contact with this air. The carbon dioxide dissolves in a film of water that covers each cell. The water is then absorbed by the cell, carrying the carbon dioxide with it.

Stomata also allow water to leave a leaf by means of evaporation. A single tree can lose 1,800 gallons of water from its leaves during a single growing season. At the same time, the tree draws in more water through the roots, along with needed minerals from the soil such as calcium, iron, and nitrates. Stomata are open during the daylight hours, when photosynthesis is taking place. They close up at night.

In this investigation, you will observe leaf stomata. You will see how they maintain a leaf's water balance and what happens when the stomata become clogged.

MATERIALS

- leaves freshly picked from several different kinds of trees
- glass microscope slides
- microscope with 100× magnification
- butyl acetate cement
- petroleum jelly
- paper toweling (for clean-up)

(Continued on next page)

103

Name _____ Date _____ Class _____

Science Activity 8 (continued)

Procedure

1. Spread a thin film of the cement on the top and underside of several leaves. Let it dry.

2. Peel the film of hardened cement from the *tops* of the leaves. Smooth it onto a glass slide and examine it under the microscope. In the Tops of Leaves section of the Data area below, sketch some of the leaf structures you see.

3. Peel the film from the *undersides* of the leaves. Examine it under the microscope. In the Undersides of Leaves section of the Data area, sketch the leaf structures you see.

4. Coat several smaller leaves with petroleum jelly. Set them aside, along with other leaves you have not coated. Observe them for a period of several days.

Data

Tops of Leaves	Undersides of Leaves

Conclusions

1. Which side of the leaves contained oval stomata? _____

2. What differences did you see in the stomata of leaves from different trees? _____

3. Compare the leaves that were coated with petroleum jelly with those that were not coated. What difference do you observe as time passes? Why? _____

4. What would happen if the stomata of the leaves of a growing tree were clogged by pollutants such as oil or ash? _____

104

Name _____ Date _____ Class _____

Science Activity 9

Hardness of Sanding Abrasives

Flint (or quartz) and garnet are two of the abrasives used to make abrasive papers for sanding. They are natural substances, known as *minerals*, that are found in the earth. Quartz is familiar to most people as the clear or white grains that make up most sandy beaches. If you look carefully among the grains of quartz on some beaches, you may also find clear reddish or-brownish grains of garnet. These two minerals have been chosen as sanding abrasives for two good reasons: they are plentiful and they are hard.

Geologists use the Mohs scale to describe the hardness of a mineral. The scale is based on the fact that a mineral can be used to scratch the surface of any other mineral softer than it is. The minerals used in the Mohs scale are shown in the table.

MATERIALS
- piece of chalk
- steel file or nail
- penny
- other materials supplied by your teacher
- garnet sandpaper

Mohs Scale

Number	Mineral
10	diamond
9	corundum
8	topaz
7	quartz
6	orthoclase
5	apatite
4	fluorite
3	calcite
2	gypsum
1	talc

Diamond can scratch any mineral, including the others on the Mohs scale. It is the hardest natural substance known. Any of the other minerals on this list can scratch minerals with a lower number on the Mohs scale.

You may not recognize many of the minerals on the Mohs scale. In this investigation, you will create a hardness scale of your own using minerals with which you are more familiar.

(Continued on next page)

105

Name _____ Date _____ Class _____

Science Activity 9 (continued)

Procedure

1. Do a scratch test using the penny, file, chalk, and any other materials your teacher supplies. Record your results for each test in the Data chart below. Which minerals will the material scratch? Which materials won't it scratch?

2. Assign the material that was scratched by all the others the lowest number of 1. Assign the material that scratched all the others the highest number of 10. Assign the other materials the appropriate number in between. Record your scale on a separate piece of paper and attach it to this page.

3. Use the piece of garnet sandpaper to scratch some of the other materials. What number would you assign it on your scale?

Data

Material	Abrasive Used	Will Scratch	Won't Scratch

Conclusions

1. The true hardness of the three suggested materials according to the Mohs scale is: file–7; penny–5; chalk–1. In this list, what Mohs hardness number would you assign the garnet sandpaper?

_____

2. What materials on the Mohs scale are probably harder than garnet? _____

_____

3. What materials on the Mohs scale are probably softer than garnet? _____

_____

4. What do you estimate the hardness of wood to be on the Mohs scale? _____

5. What must be true of any material used to sand wood? _____

Name _____ Date _____ Class _____

Science Activity 10

Identifying Simple Machines

Tools and machines help make our work easier. All of the complex machines we use today are combinations of six simple machines. Those six are the inclined plane, the wedge, the lever, the wheel and axle, the pulley, and the screw.

Although machines make our work easier, they don't actually reduce the amount of work that has to be done. Scientists use the following formula to calculate the amount of work accomplished:

Work = Force × Distance

If you increase the force, you can decrease the distance and still do the same amount of work. Machines help increase force. If you increase the distance you can decrease the force. Machines help us increase distance. For example, think of a child's teeter-totter, which is a kind of lever. Suppose you place a 100-lb. bag of sand on one end and attempt to lift it by sitting on the other end. Moving the fulcrum (pivot point) of the teeter-totter will help. The closer the fulcrum is to the sand, the easier lifting will be, because you've lengthened the lever and increased your own force.

For this activity, you will experiment with three common hand tools, identify the simple machines being used, and experiment with the advantage they provide.

MATERIALS
- 2 claw hammers with handles of different lengths
- straight-claw hammer
- common nails
- scrap wood
- saw
- hand drill
- screws

Procedure

1. If necessary, do some research on the six simple machines in a reference book or on the Internet. Identify the simple machines on which the hammer, saw, and hand drill are based and enter your conclusions on the Data chart on the next page. If you wish, you may work with the tools themselves to help you determine your answer.

2. Make a drawing in the space provided to illustrate why you have categorized the tool as that type of simple machine. For example, if you think one of the tools is a lever, you could draw it in use and label the fulcrum.

3. Hammer several nails into the scrap wood. Conduct a test to compare the different hammers in their ability to pull the nails. What influence do claw design and handle length have on the process? Describe your test in the space provided.

(Continued on next page)

Name _____ Date _____ Class _____

Science Activity 10 (continued)

Data

Tool	Type of Machine	Illustration
Hammer		
Saw		
Hand drill		

Description of Test

Conclusions

1. Which hammer was most effective at pulling nails? _____

   Why do you think this was so? _____
   _____
   _____

2. Which hammer was least effective at pulling nails?

   Why do you think this was so? _____
   _____
   _____

3. Explain in your own words why machines provide a mechanical *advantage*.
   _____
   _____
   _____
   _____

108

Name _____ Date _____ Class _____

Science Activity 11

How Solvents Work

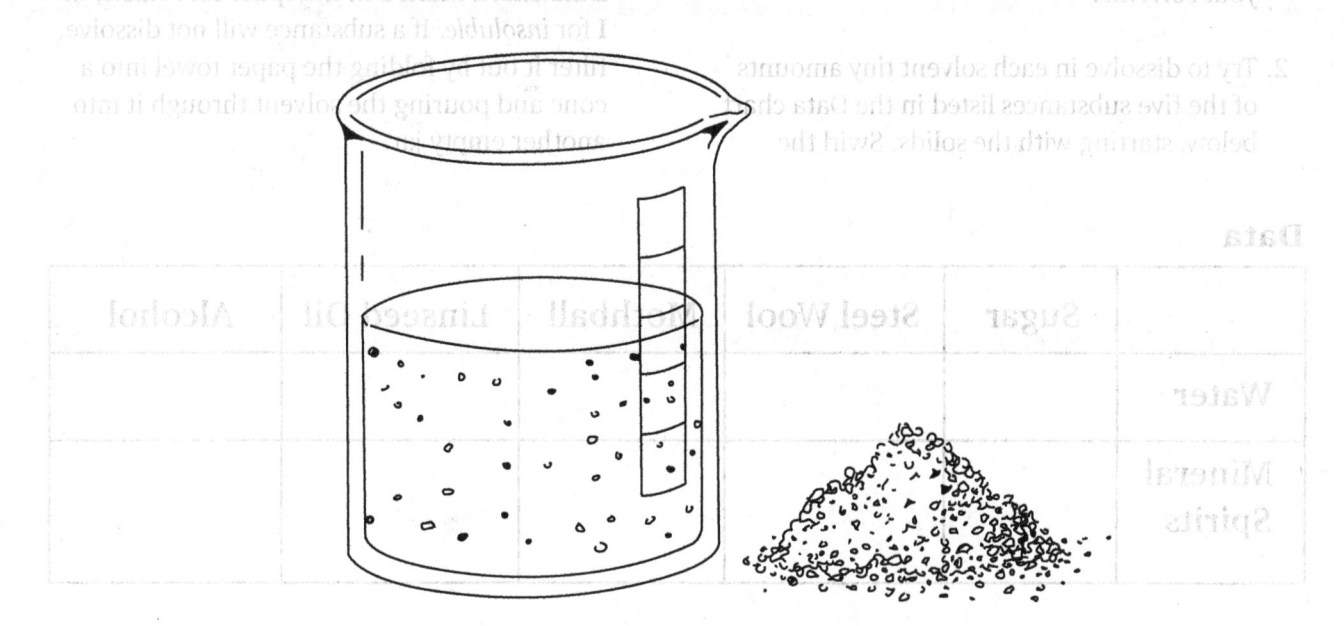

SOLVENT + SOLUTE = SOLUTION

When one substance dissolves in another, the two become completely mixed so that they appear to be a single substance. They have formed a *solution*. In a solution, the substance being dissolved is the *solute*, and the substance doing the dissolving is the *solvent*. For example, when salt dissolves in water, salt is the solute and water is the solvent.

Solids, liquids, and gases can be solvents or solutes. The air we breathe is a solution of many gases in nitrogen (gas in gas). A metal alloy is one metal dissolved in another (solid in solid). You use water as a solvent every day, for dissolving both solids and liquids.

Solvents used in woodworking are liquids. They are used to dissolve solids or other liquids. Each solvent dissolves only certain substances. Not all substances are soluble in a particular solvent. That's why you have to use the right solvent for every job. In this activity, you will investigate two different solvents: water and mineral spirits.

MATERIALS
- 4 small glass jars or beakers
- paper towels
- water
- mineral spirits
- spoonful of linseed oil
- shreds of steel wool
- mothballs
- pinch of sugar
- spoonful of rubbing alcohol

(Continued on next page)

Name _____ Date _____ Class _____

Science Activity 11 (continued)

Procedure

1. Pour water and mineral spirits into two of the jars to a depth of 1 inch (2.5 cm). These are your solvents.

2. Try to dissolve in each solvent tiny amounts of the five substances listed in the Data chart below, starting with the solids. Swirl the substance and the solvent together in the jar. Observe the results. Record your results in the Data chart. Mark **S** in the space for *soluble,* or **I** for *insoluble*. If a substance will not dissolve, filter it out by folding the paper towel into a cone and pouring the solvent through it into another empty jar.

Data

	Sugar	Steel Wool	Mothball	Linseed Oil	Alcohol
Water					
Mineral Spirits					

Conclusions

1. Which substance dissolved in neither mineral spirits nor water? _____

_____

2. Suppose you are about to refinish an old table on which the existing "solvent-based" varnish has become dark and scratched. Can you use water to remove the old varnish? Why?

_____

_____

_____

110

Name _____ Date _____ Class _____

Science Activity 12

Forces in Wood Framing

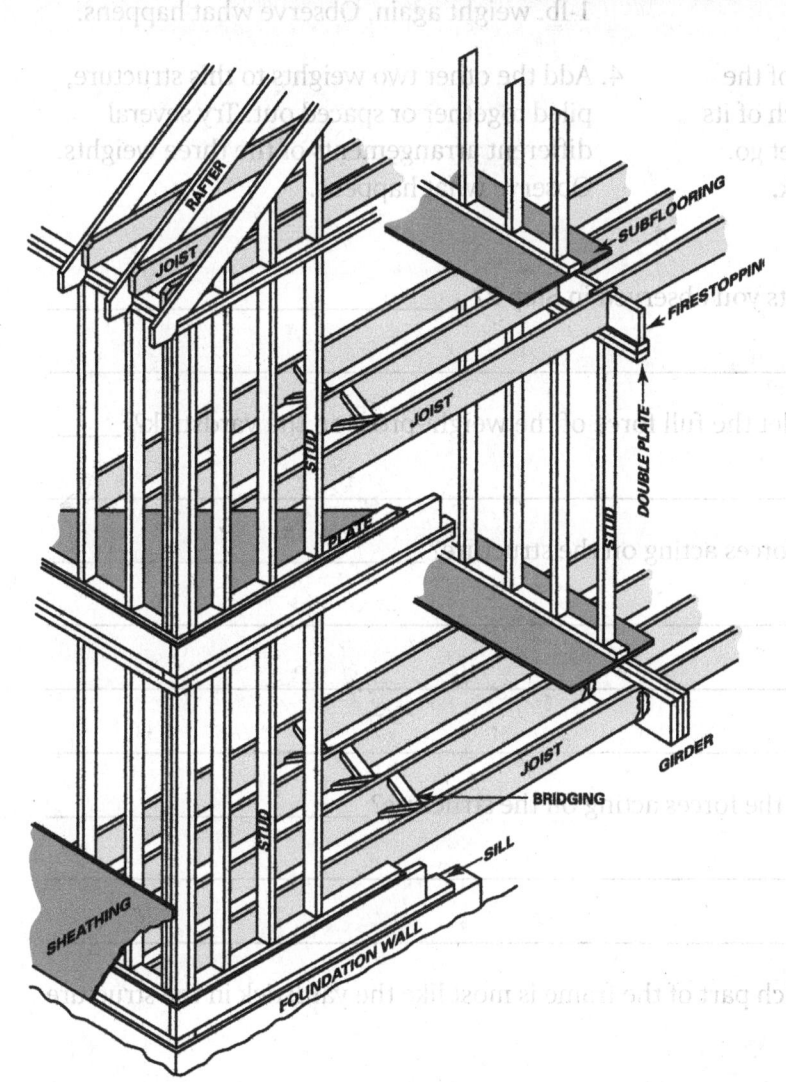

MATERIALS

- wooden yardstick
- 5 wooden blocks of the same height, about 2"×2"×6"
- 3 1-lb. weights
- masking tape

Gravity is always pulling down on you. Then why don't you sink through the floor where you are standing? It's because another force, the upward force of the floor, exactly equals the force of your weight pressing down.

In a wood-framed building, the weight of the structure and the load it will hold (such as furnishings and people) must be equaled by the upward force of wood supports. If not, the building will sag or fail. The load, or downward force, is distributed among many supports. You will study some of the forces in wood framing in this investigation.

(Continued on next page)

111

Name _____ Date _____ Class _____

Science Activity 12 (continued)

Procedure

1. Lay the yardstick flat, supported by a wooden block at each end. Tape each end of the yardstick to the top of the blocks.

2. Hold one 1-lb. weight over the center of the yardstick and lower it gently until much of its weight rests on the yardstick. Do not let go. Observe what happens to the yardstick.

3. Use the remaining three blocks to support the yardstick, spacing them out between the end blocks. Use the yardstick to support the 1-lb. weight again. Observe what happens.

4. Add the other two weights to this structure, piled together or spaced out. Try several different arrangements of the three weights. Observe what happens.

Conclusions

1. What forces were involved in the results you observed in Step 2? _____

2. What would have happened had you let the full force of the weight press on the yardstick? _____ _____

3. How did adding supports change the forces acting on the structure? _____ _____ _____

4. How did adding more weights change the forces acting on the structure? _____ _____

5. Look at the drawing on page 111. Which part of the frame is most like the yardstick in the structure you built? _____

   Which parts are like the support blocks? _____

6. What factors are taken into consideration in choosing and placing supports for a load-bearing wall in a wood-framed building? _____ _____ _____

(Continued on next page)

Name _____ Date _____ Class _____

Math Activity 1

Reading a Table

Speeds for Woodworking

Diameter of Stock	Roughing to Size	General Cutting	Finishing
Under 2"	900 to 1,300 rpm	2,400 to 2,800 rpm	3,000 to 4,000 rpm
2" to 4"	600 to 1,000 rpm	1,800 to 2,400 rpm	2,400 to 3,000 rpm
4" to 6"	600 to 800 rpm	1,200 to 1,800 rpm	1,800 to 2,400 rpm
6" to 8"	400 to 600 rpm	800 to 1,200 rpm	1,200 to 1,800 rpm
8" to 10"	300 to 400 rpm	600 to 800 rpm	900 to 1,200 rpm
Over 10"	200 to 300 rpm	300 to 600 rpm	600 to 900 rpm

In woodworking, information is often given in table form. It is important to be able to read a table correctly. The table above shows various lathe speeds used for woodworking. Use the table to complete this activity.

Questions

1. If the general cutting speed is 900 rpm, how large would you expect the diameter of the stock to be?

   _____

2. What would the minimum diameter be if the finishing speed is 1,300 rpm?

   _____

3. For what diameters would you use a speed of 200 rpm for roughing to size?

   _____

4. Which column on the table generally shows the fastest speeds for woodworking?

   _____

   The slowest? _____

5. If the diameter is 8", what is the proper finishing speed?

   _____

6. Which column shows the largest variation in woodworking speeds?

   _____

(Continued on next page)

113

Name _____ **Date** _____ **Class** _____

Math Activity 1 (continued)

7. Why does roughing to size require the slowest speeds?

_____

_____

_____

8. Approximately how many times greater is the fastest speed for general cutting than the slowest?

_____

9. What is the general cutting speed for a piece of wood that is 4" in diameter?

_____

10. On the average, how much faster is the finishing speed than the speed for roughing to size?

_____

_____

114

Name _____ Date _____ Class _____

Math Activity 2

Using the Metric System

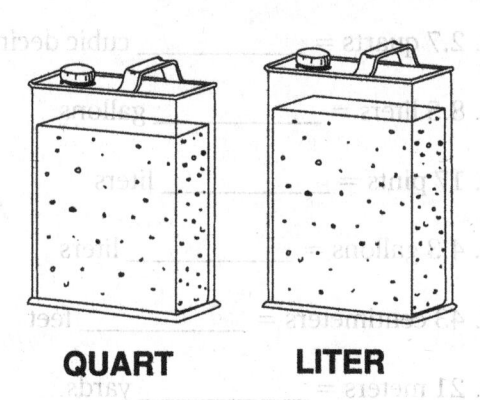

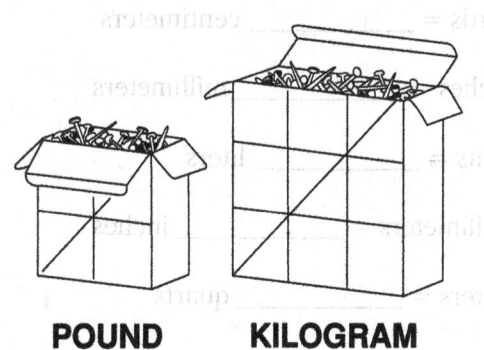

QUART   **LITER**   **POUND**   **KILOGRAM**

Woodworking requires accurate measurements so that products are made properly. For years, the United States has used the customary, or English, system of measurement. Most countries use the metric system because its units are conveniently set up in multiples of 10. The United States is making more use of the metric system than in the past. As more and more companies do business overseas, the need to understand and use the metric system will continue to grow. Use the conversion chart below to perform the conversions on page 116.

LENGTH	
Customary to Metric	**Metric to Customary**
1 inch = 25.40 millimeters (mm) 1 inch = 2.540 centimeters (cm) 1 foot = 30.480 centimeters (cm) 1 foot = 0.3048 meter (m) 1 yard = 0.9144 meter (m)	1 millimeter = 0.03937 inch 1 centimeter = 0.3937 inch 1 meter = 39.37 inches 1 meter = 3.2808 feet 1 meter = 1.0936 yards
VOLUME	
Customary to Metric	**Metric to Customary**
1 pint (liq.) = 0.473 liter (l) = 0.473 cm^3 1 quart (liq.) = 0.9463 liter (l) = 0.9463 dm^3** 1 gallon (liq.) = 3.7853 liters (l) = 3.7853 dm^3**	1 liter = 1,000 cm^3* = 0.5283 pint (liq.) 1 liter = 1 dm^3** = 1.0567 quarts (liq.) 1 liter = 1 dm^3** = 0.26417 gallon (liq.)

*cubic centimeter
**cubic decimeter

(Continued on next page)

115

Name _____ Date _____ Class _____

Math Activity 2 (continued)

Conversions

1. 2 inches = _____ centimeters
2. 4.5 feet = _____ meters
3. 3.3 yards = _____ centimeters
4. 0.3 inches = _____ millimeters
5. 1.3 pints = _____ liters
6. 2.4 millimeters = _____ inches
7. 1.75 liters = _____ quarts
8. 1.1 meters = _____ feet
9. 1.8 liters = _____ pints
10. 9.3 centimeters = _____ inches

11. 1.6 feet = _____ centimeters
12. 5.2 yards = _____ meters
13. 2.7 quarts = _____ cubic decimeters
14. 8.5 liters = _____ gallons
15. 17 pints = _____ liters
16. 4.3 gallons = _____ liters
17. 43 centimeters = _____ feet
18. 21 meters = _____ yards
19. 9.4 liters = _____ pints
20. 2.6 gallons = _____ cubic decimeters

116

Name _____ Date _____ Class _____

Math Activity 3

Reading a Working Drawing

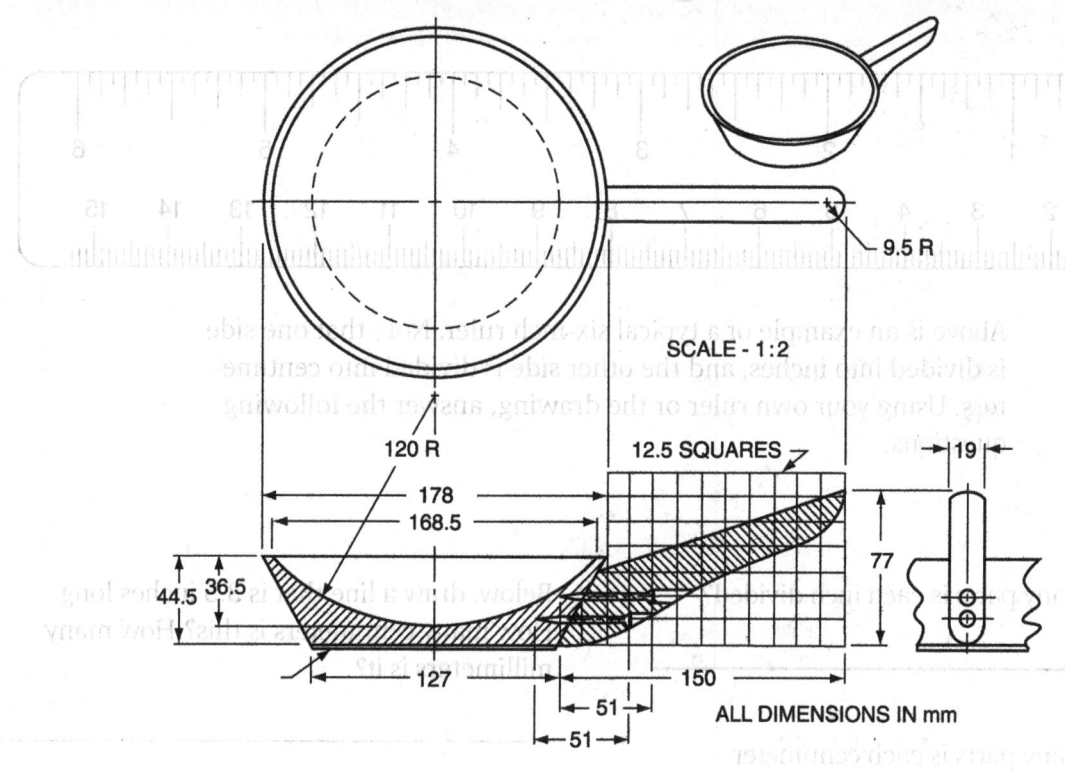

SCALE - 1:2

9.5 R
120 R
178
168.5
44.5 36.5
127
12.5 SQUARES
77
150
51
51
19
ALL DIMENSIONS IN mm

The drawing above is an example of an all-metric drawing. Refer to the drawing to answer the questions below.

Questions

1. What is the largest diameter of the turned bowl shown in the drawing?

   _____

2. What is the diameter of the bottom of the bowl?

   _____

3. What is the height of the bowl?

   _____

4. What is the radius of the arc that forms the inside of the bowl?

   _____

5. What is the thickness of the handle?

   _____

6. What is the overall length of the handle?

   _____

7. How big are the squares on which the handle design is shown?

   _____

117

Name _____ Date _____ Class _____

Math Activity 4

Using a Ruler

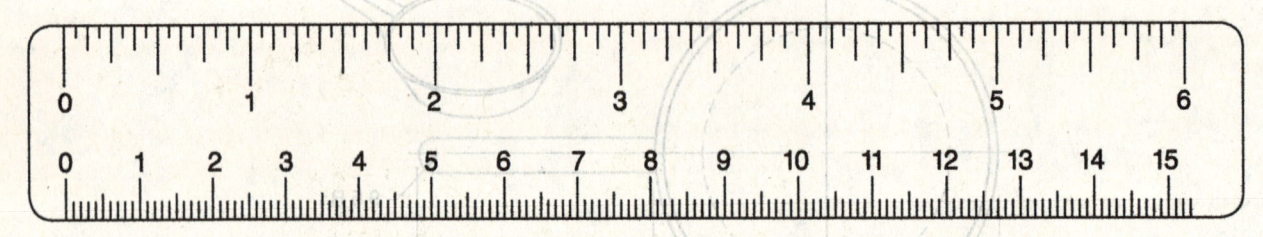

[ruler image showing 0-6 inches on top and 0-15 centimeters on bottom]

Above is an example of a typical six-inch ruler. Note that one side is divided into inches, and the other side is divided into centimeters. Using your own ruler or the drawing, answer the following questions.

Questions

1. Into how many parts is each inch divided?

   _____

2. Into how many parts is each centimeter divided? What is the name of these parts?

   _____

3. Approximately how many centimeters are in 1 inch?

   _____

4. Approximately how many millimeters are in 1 inch?

   _____

5. How many centimeters are in 1 meter?

6. Below, draw a line that is 3.5 inches long. How many centimeters is this? How many millimeters is it?

   _____

   Line:

7. Below, draw a line that is 4 centimeters long. How many inches is this?

   _____

   Line:

8. Rounding to the nearest ⅛ inch, how large is 1 centimeter in fractional inches?

   _____

9. How many millimeters are in 6 inches?

   _____

118

Name _____ Date _____ Class _____

Math Activity 5

Fractions and Decimals

Accuracy with fractions and decimals is essential in all phases of woodworking. The following exercises will test your skills with these kinds of numbers.

Fraction-to-Decimal Conversion
Change the following fractions to decimals. Record each answer to four decimal places.

1. 3/4 _____
2. 1/8 _____
3. 7/16 _____
4. 3/32 _____
5. 2/9 _____

6. 4/7 _____
7. 5/6 _____
8. 11/64 _____
9. 5/4 _____
10. 21/16 _____

Adding Fractions
Add the following fractions and reduce the answers to the lowest terms.

11. 1/2 + 3/16 = _____
12. 3/32 + 2/64 = _____
13. 3/5 + 1/6 = _____
14. 1/4 + 1/3 = _____
15. 5/8 + 3/4 = _____

16. 2/3 + 5/8 = _____
17. 1/32 + 3/16 = _____
18. 4/9 + 3/4 = _____
19. 1/16 + 2/3 = _____
20. 1/5 + 1/4 = _____

Multiplying Fractions
Multiply the following fractions and reduce the answers to the lowest terms.

21. 3/8 × 5/6 = _____
22. 1 1/2 × 2/3 = _____
23. 8 × 5/16 = _____
24. 4 3/4 × 2 1/5 = _____
25. 5/64 × 4/5 = _____

26. 7/8 × 3/4 = _____
27. 12 × 3 5/6 = _____
28. 1/2 × 3/4 = _____
29. 7/46 × 23/28 = _____
30. 1/25 × 5/7 = _____

(Continued on next page)

119

Name _____ Date _____ Class _____

Math Activity 5 (continued)

Dividing Fractions
Divide the following fractions and reduce the answers to the lowest terms.

31. $3/4 \div 1/8 =$ _____

32. $2\ 1/3 \div 5/6 =$ _____

33. $2/5 \div 2 =$ _____

34. $1\ 1/2 \div 3\ 3/8 =$ _____

35. $1/4 \div 8 =$ _____

36. $4\ 1/4 \div 5/16 =$ _____

37. $3/4 \div 1/32 =$ _____

38. $11/16 \div 1/2 =$ _____

39. $138\ 2/3 \div 4 =$ _____

40. $15 \div 2/3 =$ _____

Subtracting Fractions
Subtract the following fractions and reduce the answers to the lowest terms.

41. $5/8 - 1/6 =$ _____

42. $19/32 - 15/32 =$ _____

43. $3\ 1/4 - 2/3 =$ _____

44. $1/2 - 1/16 =$ _____

45. $4\ 2/5 - 2\ 1/2 =$ _____

46. $7/8 - 3/4 =$ _____

47. $2\ 1/2 - 9/16 =$ _____

48. $5\ 1/3 - 2\ 1/2 =$ _____

49. $1\ 7/8 - 11/64 =$ _____

50. $10\ 5/8 - 6\ 3/4 =$ _____

(Continued on next page)

120

Name _____ Date _____ Class _____

Math Activity 6

Figuring Percentages

The ability to figure percentages is an important skill. First, remember that percents are always based on 100. To find the percent of a whole number, change the parts of the problem to decimals and solve. For example, to find 15% of 20, change 15% to 0.15 and multiply. The answer is 3 (20 × 0.15 = 3). The questions and problems in this activity will help you practice working with percentages.

Questions

1. 40% of 80 = _____

2. 25% of 75 = _____

3. 15% of 60 = _____

4. 52% of 100 = _____

5. 52% of 50 = _____

6. 30 is 25% of _____

7. 18 is 5% of _____

8. 53 is 53% of _____

9. 8 is 30% of _____

10. 15 is 150% of _____

Problems

11. If a 2" line is changed to 2¼", by what percentage has it changed?

    _____

12. Suppose the average weekly wage of a carpenter was $110 in 1960. By 1990, the average carpenter was earning $495. What is the percentage of increase?

13. In 1980, 20,000 new public housing units were started. In 1997, only 2,000 new units were begun. What is the percentage of decrease in public housing?

    _____

14. If a piece of wood 4" long is shortened to 2.6", by what percentage has it changed?

    _____

121

Name _____ Date _____ Class _____

Math Activity 7

Figuring Square and Board Feet

Bill of Materials

Stock: Pine

IMPORTANT: All dimensions listed below are finished size.

No. of Pieces	Part Name	Thickness	Width	Length
1	Side	½"	12½"	28"
1	Side	½"	12¼"	28"
1	Shelf 1	½"	11½"	11½"
1	Shelf 2	½"	9¼"	9¼"
1	Shelf 3	½"	7¼"	7¼"
1	Shelf 4	½"	5⅜"	5⅜"
1	Shelf 5	½"	3¾"	3¾"
2	Flat Metal Hangers			
24	1¼"-8 Wood Screws			

The bill of materials shown above is for a corner cabinet with five tiered shelves. Use the bill of materials to answer the following questions. For all calculations, assume 10% extra for rough size, and remember that stock less than 1 inch is figured as 1 inch. Figure your answers to four decimal places.

Questions

1. How many board feet are needed for the two sides of this project?

2. How many board feet are needed for shelf 1?

3. How many board feet are needed for all of the shelves combined?

4. What is the minimum number of board feet of pine you should buy for this project?

5. If you decided to build the cabinet without shelves 2 and 4, how would that change the minimum number of board feet of pine you would need?

122

Name _____ Date _____ Class _____

Math Activity 8

Wood Screw Sizes

Wood screws come in various sizes from 0 to 24. The smallest size is 0, and that screw has a diameter of 0.060 inch. Each succeeding screw has a diameter that is 0.013 inch larger than the previous one. Refer to Table 15-A on page 229 of Wood Technology & Processes to answer the following questions.

Questions

1. How much larger is the diameter of a number 24 screw than the diameter of a number 1 screw?

   _____

2. How long is the longest number 16 screw?

   _____

3. What screw size is almost three times the diameter of a number 1 screw?

   _____

4. How many times longer is the longest number 20 screw than the number 1 screw?

   _____

5. How much bigger is the diameter of a number 16 screw than the diameter of a number 9 screw?

   _____

6. Which is larger for the number 24 screw, the pilot hole for soft wood or for hard wood?

   _____

   By how much? _____

7. What is the difference between the largest and smallest pilot holes for soft wood?

   _____

   For hard wood? _____

8. How much do the pilot holes for soft and hard wood differ for the number 6 screw?

   _____

9. For which number screws are the pilot holes for hard and soft wood the same size?

   _____

10. Compare the soft wood pilot holes for a number 2 and a number 18 screw. Which is larger?

    _____

    By how many times? _____

123

Name _____ Date _____ Class _____

Math Activity 9

Finishing Coverage

CHEST OF DRAWERS

Dimensions: 37" wide, 18" deep, 30" tall

Wood furniture is usually finished. Finishing materials include stains, fillers, and lacquers. In this exercise, refer to the drawing to answer the questions.

Assume that the typical coverages for the finishing materials are:

Stain: 350 square feet per gallon
Filler: 225 square feet per gallon
Lacquer: 175 square feet per gallon

Questions

1. What are the measurements of the following parts of the chest of drawers? Give your answers in square inches.

   a. Front _____

   b. Top _____

   c. Left side _____

   d. Right side _____

2. Add the numbers in question 1 and divide by 144. What does this number represent?

   _____

3. Using the result obtained in question 2, determine exactly how many of these chests can be covered by:

   a. One gallon of stain _____

   b. One gallon of filler _____

   c. One gallon of lacquer _____

4. What is the total number of square feet in the chest, including the back and bottom?

   _____

5. How many square inches can one gallon of filler cover? _____

124

Name _____ Date _____ Class _____

Math Activity 10

Cost of Electricity for Woodworking Machines

Many woodworking machines require electricity to operate. Their operating cost is determined by how much electricity they use. Electricity is measured in watts. One thousand watts equal one kilowatt. Utility companies usually charge by the number of kilowatts used in one hour.

To find out how much electricity a machine uses, first find out how many hours it is used each month. Then figure the watt-hours. Watt-hours are calculated using the formula:

watt-hours = hours used × power in watts

Next, turn watt-hours into kilowatt-hours. To do this, use the formula:

$$kWh = \frac{\text{watt-hours}}{1{,}000}$$

Read each of the following scenarios. Then use these formulas to help answer the questions.

Scenario 1

A contractor's circular saw uses an average of 925 watts of power. She operates the saw 7 hours a day and 22 days a month. The power company charges $0.14 per kWh.

Questions

1. For how many hours each month does the contractor use the saw?

2. How many watt-hours does the circular saw use?

3. How many kilowatt-hours does the saw use each month?

4. What is the monthly cost to operate the power saw?

5. What is the daily cost?

6. What is the yearly cost?

(Continued on next page)

125

Name _____ Date _____ Class _____

Math Activity 10 (continued)

Scenario 2

A drill press is operated for four hours each day and 15 days each month. It uses an average of 1,100 watts of power. The cost for this power is $0.09 per kWh.

Questions

1. How many hours each month is the drill press used?

2. How many watt-hours does the drill press use each month?

3. How many kilowatt-hours does the press use each month?

4. How many kilowatt-hours does the press use each year?

5. How much does it cost to operate the drill press each month?

6. How much does it cost to operate the drill press each day?

7. What is the yearly cost?

8. If the cost of the power went up to $0.13 per kWh, how much more per year would it cost to run the drill press?

126

Name _____ Date _____ Class _____

Workshop Safety

Short Answer

Directions: Answer each question in the space provided.

1. In general, what is the significance of the color red on machinery and around the workshop?
   _____
   _____

2. Name three common hazards in the workshop.
   _____
   _____

3. List two common causes of accidents in the workshop and explain how to avoid them.
   _____
   _____
   _____
   _____

4. Explain why ground-fault circuit interrupters should be used in the workshop.
   _____
   _____
   _____

5. List three types of hazardous materials commonly found in the workshop and describe the potential hazard each presents.
   _____
   _____
   _____

6. What is a Material Safety Data Sheet? How should you use it?
   _____
   _____
   _____

(Continued on next page)

Name _____ Date _____ Class _____

Workshop Safety (continued)

7. If you are unsure how to dispose of a hazardous chemical, how can you find out?
   _____
   _____

8. Describe appropriate dress for the woodworking workshop.
   _____
   _____
   _____

9. When should you wear safety glasses or goggles?
   _____
   _____

10. Describe the dangers associated with noise levels generated by power tools.
    _____
    _____

11. What is the best way to protect yourself from harmful dust and vapors?
    _____
    _____

12. When should you wear gloves in the workshop? When shouldn't you?
    _____
    _____

13. What safety factors should you take into consideration when setting up a workshop?
    _____
    _____
    _____

14. What should you do if a coworker experiences an electrical shock and cannot let go of the source of the shock?
    _____
    _____
    _____

(Continued on next page)

Name _____ Date _____ Class _____

Fire Safety

Short Answer

Directions: Answer each question in the space provided.

1. How often should a smoke detector be tested to be sure the battery remains effective?

2. What type of fire extinguisher is effective against wood, oil, solvent, chemical, and electrical fires?

3. What elements should a fire safety plan include?

4. How should you store finishing products and solvents in the workshop?

5. Explain the possible consequence of replacing a fuse with one of higher amperage.

6. Where should you store oily or paint-covered rags? Why?

7. Why is it important to check plugs and power cords regularly?

(Continued on next page)

129

Name _____ Date _____ Class _____

Fire Safety (continued)

8. Why should a woodworking shop be well ventilated?

9. Name three things regarding fire safety that everyone who works in a woodworking shop should know about that particular workshop.

10. Briefly describe the three basic ways to control a fire.

11. What should you do if a person's clothing catches on fire?

12. When should you **not** throw water on a fire?

13. Why is it **not** a good idea to use multiple power strips on a single circuit so that you can plug in a large number of electric tools?

14. Explain the importance of cleaning the workshop frequently to remove sawdust, wood chips, and any solvents that may have spilled.

(Continued on next page)

Name _____ Date _____ Class _____

Hand/Portable Power Tool Safety

Short Answer

Directions: Answer each question in the space provided.

1. Name two precautions you should take when using handsaws.

2. What should you do before making any adjustments to a portable power saw?

3. When using a portable power saw, in which direction should you cut? Why?

4. What things should you check before plugging in a portable power saw?

5. Why is it important to wear safety glasses while using a hammer?

6. What should you do before installing drill bits or other tools on a cordless power drill?

7. Name two reasons eye protection is important while using a portable power drill.

(Continued on next page)

131

Name _____ Date _____ Class _____

Hand/Portable Power Tool Safety (continued)

8. Describe the procedure for installing a drill bit in a portable drill safely.
_____
_____
_____
_____
_____

9. What are two precautions you should take when using a hand or portable power sander?
_____
_____
_____
_____

10. Why is it necessary to use clamps when working with a biscuit joiner?
_____
_____
_____
_____

11. How should you hold a biscuit joiner while cutting biscuits?
_____
_____
_____

12. Name two precautions you should take when using a hand plane.
_____
_____
_____

13. Where should your hands be when you are using a chisel?
_____
_____
_____

14. Why is it important to use hand tools only for the jobs they were designed to do?
_____
_____
_____

(Continued on next page)

Name _____ Date _____ Class _____

Planer Safety

Short Answer

Directions: Answer each question in the space provided.

1. For what things should you check a board before planing it?
   _____
   _____
   _____
   _____

2. What is the **minimum** length a board should be for planing?
   _____
   _____

3. Can you plane a warped board? Explain.
   _____
   _____
   _____
   _____

4. Where should you stand while planing a board?
   _____
   _____

5. If you have several boards of different thicknesses to be planed, how should you proceed?
   _____
   _____
   _____
   _____

(Continued on next page)

133

Name _____ Date _____ Class _____

Planer Safety (continued)

6. Why should you never look into the planer while the board is passing through?
   _____
   _____

7. What should you do if a board sticks in the planer?
   _____
   _____
   _____
   _____

8. How should you support a long board as it leaves the planer?
   _____
   _____
   _____

9. What does tailing off involve? Describe how to tail off properly and safely.
   _____
   _____
   _____
   _____
   _____

10. What should you do before making any adjustments or cleaning the planer?
    _____
    _____
    _____

Name _____ Date _____ Class _____

Jointer Safety

Short Answer

Directions: Answer each question in the space provided.

1. What personal protective equipment should you wear while jointing?

2. What should you do before you adjust the tables for the depth of the cut?

3. What are the **minimum** dimensions of stock to be cut in a jointer?

4. Describe how to perform a face planing (surfacing) operation safely on the jointer.

5. Where should you stand while operating a jointer?

(Continued on next page)

Name _____ Date _____ Class _____

Jointer Safety (continued)

6. What is the **only** exception to the rule that the guard must be in place over the knives while the jointer is being operated? What should you do instead?

_____
_____
_____
_____
_____

7. In what direction should you feed stock into a jointer?

_____
_____

8. How should you hold the workpiece while performing an edge jointing operation?

_____
_____

9. Describe the safest way to cut a bevel or chamfer on an edge using a jointer.

_____
_____
_____
_____
_____

10. What margin of safety should you use when jointing stock? Where should your hands be placed?

_____
_____
_____

11. What is the deepest cut you should make on a jointer? Why?

_____
_____
_____
_____

(Continued on next page)

Name _____ Date _____ Class _____

Table Saw Safety

Short Answer

Directions: Answer each question in the space provided.

1. What is the best way to be sure the saw will not accidentally start while you are changing saw blades?

   _____
   _____

2. What should you do before making any adjustments to a table saw?

   _____

3. How much higher than the thickness of the stock should you raise the saw blade?

   _____

4. What is the purpose of the antikickback pawls?

   _____
   _____

5. Is it ever acceptable practice to operate a table saw without the saw guard in place? Explain.

   _____
   _____

6. Where should you stand while operating a table saw?

   _____

7. When is it necessary to use a pushstick?

   _____
   _____

8. What is the **minimum** distance you should maintain between your fingers and the saw blade?

   _____

(Continued on next page)

137

Table Saw Safety (continued)

9. Is it ever acceptable practice to reach over a saw blade? Explain.

10. Describe the proper procedure for clearing away scraps that are close to the saw blade.

11. What type of fence should you use for a ripping operation?

12. Describe the requirements for stock that will be ripped. What characteristics should you look for?

13. Describe the proper procedure for tailing off when a board is being ripped.

14. What should be the first step for all crosscutting operations?

Name _____ Date _____ Class _____

Radial-Arm Saw Safety

Short Answer

Directions: Answer each question in the space provided.

1. What precautions should you take when choosing a saw blade for the radial-arm saw?
   _____
   _____
   _____

2. How should you mount the saw blade on the arbor of a radial-arm saw?
   _____
   _____
   _____

3. What two things should you do before starting the motor?
   _____
   _____

4. What personal protective equipment should you wear while using a radial-arm saw?
   _____
   _____

5. How long should you wait after starting the saw before you begin cutting the stock?
   _____
   _____

6. Where should your hands be while cutting stock on a radial-arm saw? What margin of safety should you maintain?
   _____
   _____
   _____

(Continued on next page)

Name _____ Date _____ Class _____

Radial-Arm Saw Safety (continued)

7. How should you handle stock during crosscutting operations?

8. How should you feed stock into the radial-arm saw blade when ripping?

9. What should you do before making any adjustments to the radial-arm saw?

10. What should you do after completing the cut but before removing stock from the table?

(Continued on next page)

Band Saw Safety

Short Answer

Directions: Answer each question in the space provided.

1. What personal protective equipment should you wear while using a band saw?

2. What rule should you follow when selecting the blade to use for a specific operation?

3. Describe how to check the blade before operating a band saw.

4. How should the wheels turn as viewed from the front of the band saw? How can you tell?

5. How should the upper guide assembly be adjusted in relation to the stock?

6. How should you hold the stock while cutting it with a band saw?

(Continued on next page)

Name _____ Date _____ Class _____

Band Saw Safety (continued)

7. Explain how to cut cylindrical (round) stock.
_____
_____
_____

8. Where should your fingers be while you are cutting with a band saw? What margin of safety should you maintain?
_____
_____

9. How should you feed stock into the band saw?
_____
_____

10. Explain the procedure for backing the saw blade out of a long cut safely.
_____
_____
_____
_____

11. How should you clear away scraps that are close to the saw blade?
_____
_____

12. What should you do before making any adjustments to a band saw?
_____
_____
_____
_____

(Continued on next page)

Name _____ Date _____ Class _____

Sliding Compound Miter Saw Safety

Short Answer

Directions: Answer each question in the space provided.

1. What personal protective equipment should you wear while using a sliding compound miter saw?

   _____
   _____

2. How should a sliding compound miter saw be set up in the workshop for safety?

   _____
   _____

3. What three things should you check the blade for before plugging in the machine?

   _____
   _____
   _____
   _____
   _____
   _____
   _____

4. Why is it necessary to hold the handle tightly when switching the machine on?

   _____
   _____
   _____

5. Where should your hands be when you are using the sliding compound miter saw?

   _____
   _____

(Continued on next page)

Name _____ Date _____ Class _____

Sliding Compound Miter Saw Safety (continued)

6. Explain the use of hold-down clamps with a sliding compound miter saw. How should the clamps be positioned?

_____
_____
_____
_____

7. How long should you wait after turning the saw on before you begin cutting?

_____
_____

8. What should you do after turning off the sliding compound miter saw but before removing the workpiece?

_____
_____
_____
_____

9. What should you do before inspecting or servicing a sliding compound miter saw?

_____
_____
_____

144

Name _____ Date _____ Class _____

Scroll Saw Safety

Short Answer

Directions: Answer each question in the space provided.

1. What personal protective equipment should you wear while using a scroll saw?

   _____
   _____
   _____

2. How should the blade be installed on a scroll saw?

   _____
   _____
   _____

3. Explain how to check the belt and blade before operating a scroll saw. What safety precautions should you take?

   _____
   _____
   _____
   _____
   _____
   _____
   _____

4. How can you prevent the workpiece from moving up and down with the blade?

   _____
   _____
   _____

(Continued on next page)

145

Name _____ Date _____ Class _____

Scroll Saw Safety (continued)

5. Where should your fingers be while you are cutting stock with a scroll saw?
_____
_____
_____

6. What margin of safety should you maintain?
_____
_____
_____

7. Explain how to cut cylindrical (round) stock using a scroll saw.
_____
_____
_____
_____
_____

8. What should you do before making any adjustments to a scroll saw?
_____
_____
_____

(Continued on next page)

146

Name _____ Date _____ Class _____

Drill Press Safety

Short Answer

Directions: Answer each question in the space provided.

1. What personal protective equipment should you wear while using a drill press?

2. What type of shank must a bit have to be used with a drill press?

3. What is a safe speed for operating a drill press? Explain.

4. How should you secure the workpiece while operating the drill press at high speeds?

5. Explain how to install a drill bit in the chuck of a drill press safely.

6. How can you protect the vise or table under the workpiece?

(Continued on next page)

147

Name _____ Date _____ Class _____

Drill Press Safety (continued)

7. Can you use a drill press to drill cylindrical (round) stock? Explain.
   _____
   _____
   _____

8. What margin of safety should you maintain while using a drill press?
   _____
   _____
   _____

9. Explain how to remove chips and shavings from around the blade safely.
   _____
   _____
   _____
   _____

10. What should you do before performing special operations such as shaping or sanding using the drill press?
    _____
    _____
    _____

11. Explain how to change the speed safely on a variable speed machine and on a step pulley machine.
    _____
    _____
    _____
    _____
    _____
    _____

(Continued on next page)

Name _____ Date _____ Class _____

Router Safety

Short Answer

Directions: Answer each question in the space provided.

1. What personal protective equipment should you wear while using a router?

2. Explain how to select and install a router bit safely.

3. How should the workpiece be held during a routing operation?

4. Where should your hands be while you are using the router?

(Continued on next page)

Name _____ Date _____ Class _____

Router Safety (continued)

5. Explain how to feed the router into stock safely.
_____
_____
_____
_____
_____

6. Describe how to gauge whether your router is running at a safe, appropriate speed.
_____
_____
_____
_____
_____
_____
_____

7. Explain how to handle a router safely when turning it off.
_____
_____
_____
_____
_____
_____
_____

(Continued on next page)

Name _____ Date _____ Class _____

Sander Safety

Short Answer

Directions: Answer each question in the space provided.

1. What personal protective equipment should you wear while using a sander?

2. Describe the safety checks you should perform before using a power sander.

3. How can you control the position of the workpiece safely?

4. Explain how to secure small or irregularly shaped pieces for sanding.

5. When should you **not** use a power sander to shape parts?

(Continued on next page)

151

Name _____ Date _____ Class _____

Sander Safety (continued)

6. How can you prevent burning stock due to friction?
   _____
   _____

7. How should you support the workpiece when using a stationary belt sander to sand the end grain of narrow pieces?
   _____
   _____
   _____

8. Explain how to feed stock into a stationary belt sander. Why is this important?
   _____
   _____
   _____
   _____
   _____
   _____
   _____

9. What items should you check before using a stationary disc sander?
   _____
   _____
   _____
   _____
   _____
   _____

10. On what portion of a stationary disc sander should you sand stock?
    _____
    _____
    _____
    _____

Name _____ Date _____ Class _____

Lathe Safety

Short Answer

Directions: Answer each question in the space provided.

1. What personal protective equipment should you wear while using a lathe?

2. What should you check the wood for before placing it in the lathe?

3. Why is it important to allow glued-up work to cure at least 24 hours before turning it on a lathe?

4. What should you do to the tailstock before operating the lathe?

5. What is end play, and how should you check for it before operating the lathe?

6. Where should the tool rest be placed? Why?

(Continued on next page)

153

Lathe Safety (continued)

7. How can you make sure rough stock will clear the tool rest?

8. Explain the procedure for adjusting the tool rest safely.

9. At what speed should you start all turning operations?

10. How should you determine the speed at which to turn stock? Explain.

11. Why is it necessary to hold all turning tools firmly in both hands?

12. What should you do before checking the diameter of the stock with calipers?

(Continued on next page)

Name _____ Date _____ Class _____

Adhesives/Hazardous Materials Safety

Short Answer

Directions: Answer each question in the space provided.

1. What personal protective equipment should you wear while using hazardous materials?

2. Why is it important to wear a charcoal-filtered respirator when using certain types of adhesives?

3. Name two reasons it is important to work in a well-ventilated area when you are using adhesives.

4. What special precautions should you take when working with a glue gun?

5. What are hazardous materials? Give at least three examples that are commonly found in the workshop.

(Continued on next page)

Name _____ Date _____ Class _____

Adhesives/Hazardous Materials Safety (continued)

6. Some solvents can cause physical illness. Explain how solvents used in paint thinners and finishes can affect you if you do not wear the proper protective equipment.

_____
_____
_____
_____

7. Explain how harmful solvents can be absorbed into the human body.

_____
_____
_____
_____

8. What chemicals should you avoid, if possible, when you choose solvents and other products that are commonly needed in the woodworking shop?

_____
_____
_____
_____
_____
_____

9. What precautions should you take when working with South American mahogany or Western red cedar? Why?

_____
_____
_____
_____

10. How should hazardous wastes be stored in the woodworking shop?

_____
_____
_____
_____

(Continued on next page)

156

Name _____ Date _____ Class _____

Handling Finishes Safely

Short Answer

Directions: Answer each question in the space provided.

1. What personal protective equipment should you wear while handling finishes?

   _____
   _____
   _____
   _____
   _____
   _____

2. Briefly describe how to handle and store finishes and solvents in the woodworking shop.

   _____
   _____
   _____
   _____
   _____
   _____

3. How should you dispose of rags and disposable gloves after handling finishes?

   _____
   _____

4. Why is it important to keep the entire finishing area clean and free of spills?

   _____
   _____
   _____
   _____

(Continued on next page)

Name _____ Date _____ Class _____

Handling Finishes Safely (continued)

5. Describe a safe setup for spraying finishes onto woodworking projects.

_____

_____

_____

_____

6. Explain why opened finishing materials should never be left unattended.

_____

_____

_____

_____

_____

7. Why should the proper types of fire extinguishers always be available in a woodworking finishing room?

_____

_____

_____

_____

8. Why is it a good idea to perform all finishing processes in a separate room, away from all power tools and machines?

_____

_____

_____

_____

_____

Name _____ Date _____ Class _____

Personal Protective Equipment

Short Answer

Directions: Answer each question in the space provided.

1. What do the initials PPE stand for?

   _____
   _____
   _____

2. Which three types of protection should you find out about before you begin a particular job?

   _____
   _____
   _____
   _____

3. What are the main differences between disposable and reusable earplugs?

   _____
   _____
   _____
   _____

4. For what type of work are safety goggles especially useful?

   _____
   _____
   _____

5. Which type of face mask is not recommended for wood shops?

   _____
   _____
   _____

(Continued on next page)

Name _____ Date _____ Class _____

Personal Protective Equipment (continued)

6. For which tasks is a disposable respirator recommended?

_____
_____
_____
_____
_____

7. For which tasks is a reusable respirator recommended?

_____
_____
_____
_____
_____
_____
_____
_____
_____
_____
_____
_____
_____
_____
_____
_____

160

Name _____ Date _____ Class _____

Woodworking Activity 1

Making a Miter Box

SAFETY FIRST → Before doing this activity, make sure you understand how to use all the tools and materials safely.

MATERIALS
- straightedge
- hand or power saw
- ¾" hardwood stock
- wood screws
- glue

If commercial miter boxes are not available, you can easily make your own out of wood. When your miter box is paired with a hacksaw or fine-tooth crosscut saw, you can use it to make cuts at 45- or 90-degree angles. Refer to the illustrations as you work.

45°-ANGLE CUTS   90°-ANGLE CUTS
45°-ANGLE CUTS
DISTANCE EQUAL TO ½ WIDTH OF BOX

Procedure

1. Cut three pieces of wood to measure approximately 4" × 6" × ¾".
2. Fasten the pieces together with glue or screws to form a U-shaped box.
3. Lay out one cut in the center of each side of the box. The cut should be drawn at a right angle to that side.
4. Measure the width of the box.
5. At a distance equal to half the width of the box, lay out cuts on each side of both center cuts. These cuts should be drawn at a 45-degree angle, one slanting to the right and the other to the left (for cutting from either the right or the left).
6. Make all the cuts in the sides of the box.

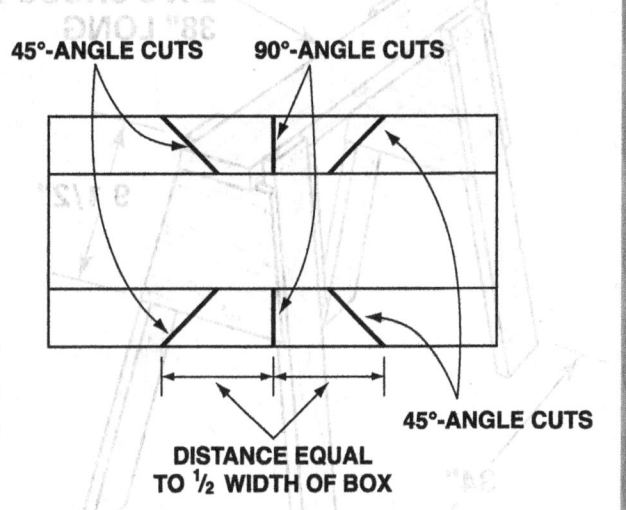

When you use the miter box, be sure the line to be cut is directly under the saw teeth. Hold the work firmly against the back of the box. Start the cut with a careful backstroke.

161

Name _____ Date _____ Class _____

Woodworking Activity 2

Building Your Own Sawhorses

SAFETY FIRST → Before doing this activity, make sure you understand how to use all the tools and materials safely.

Although sawhorses are readily available at construction supply centers, it is less expensive to make your own. This activity features two types, a standard model and a folding model.

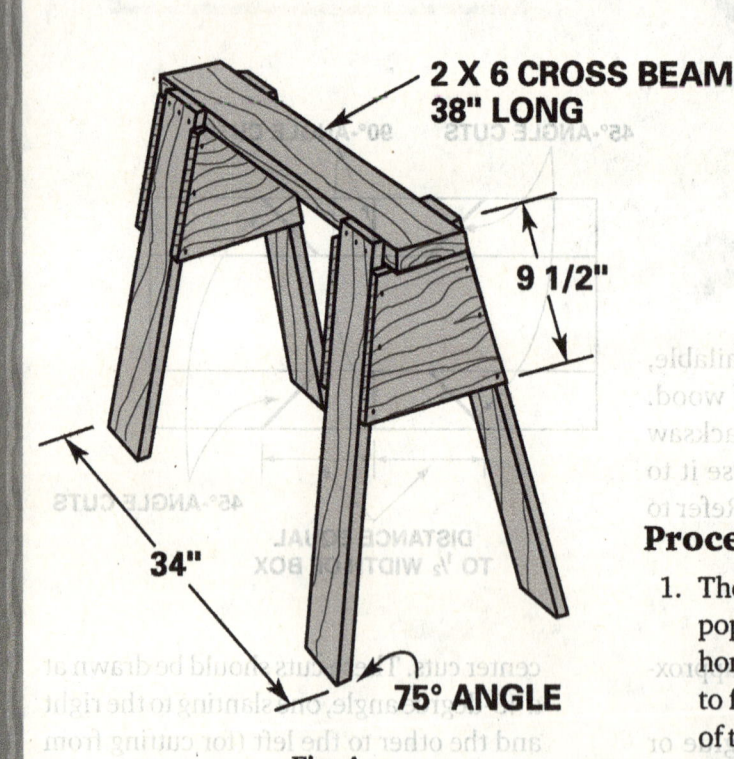

MATERIALS
- 2×6
- 2×4s
- 1×4s
- 3" butt hinges
- wood screws
- glue

2 X 6 CROSS BEAM 38" LONG

9 1/2"

34"

75° ANGLE

Fig. A

Procedure

1. The sawhorse in Fig. A is one of the most popular designs and one of the easiest sawhorses you can build. The materials are easy to find and fairly inexpensive. Both edges of the top are ripped to a 75-degree angle. The bevel on the legs is also 75 degrees. Extra gussets attached to the inside of the legs give added strength. Use construction adhesive and wood screws (not drywall screws) to attach the legs.

(Continued on next page)

162

Name _____ Date _____ Class _____

Woodworking Activity 2 (continued)

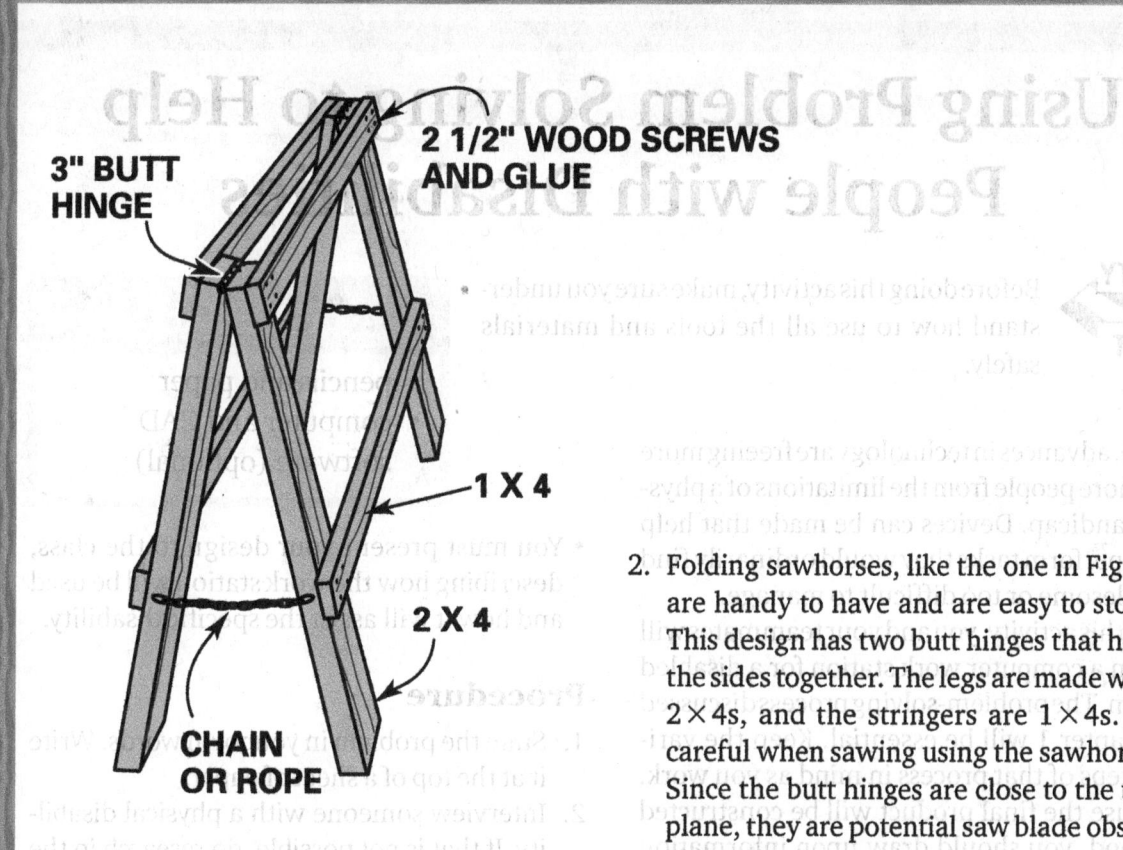

Fig. B labels:
- 3" BUTT HINGE
- 2 1/2" WOOD SCREWS AND GLUE
- 1 × 4
- 2 × 4
- CHAIN OR ROPE

2. Folding sawhorses, like the one in Fig. B, are handy to have and are easy to store. This design has two butt hinges that hold the sides together. The legs are made with 2×4s, and the stringers are 1×4s. Be careful when sawing using the sawhorse. Since the butt hinges are close to the top plane, they are potential saw blade obstacles. Also be sure the legs are spread out for stability. Rope or chain can be used to keep the legs from spreading too much.

163

Name _____ Date _____ Class _____

Woodworking Activity 3

Using Problem Solving to Help People with Disabilities

SAFETY FIRST Before doing this activity, make sure you understand how to use all the tools and materials safely.

MATERIALS
- pencils and paper
- computer and CAD software (optional)

Today, advances in technology are freeing more and more people from the limitations of a physical handicap. Devices can be made that help them perform tasks they would ordinarily find troublesome or too difficult to manage.

For this activity, you and your teammates will design a computer work station for a disabled person. The problem-solving process discussed in Chapter 1 will be essential. Keep the various steps of that process in mind as you work. Because the final product will be constructed of wood, you should draw upon information you have learned in class.

Your work station design can be intended for any type of physical disability. It can be a completely original idea or a modification of an existing work station design.

Specifications

Your design will have to meet certain standards. Read the following specifications carefully before you begin:

- Your design must help a person with a specific physical disability.
- The work station must be safe to use.
- The work station must be cost effective to make.
- You must hand in a written description of your design and identify the type of disability for which it is intended.
- You must provide drawings that clearly illustrate the design.
- You must present your design to the class, describing how the work station will be used and how it will assist the specific disability.

Procedure

1. State the problem in your own words. Write it at the top of a sheet of paper.
2. Interview someone with a physical disability. If that is not possible, do research in the library or on the Internet to learn about living with disabilities.
3. Brainstorm with your teammates to come up with as many design ideas as you can.
4. Select two or three ideas to develop. Be sure they meet all the specifications for this activity.
5. Draw rough sketches of the selected ideas. Make notes about each design to describe how it will work.
6. Consider the advantages and disadvantages of each design, including safety and cost. Select the one that seems the most promising.
7. Create a finished drawing of the work station and write your descriptive paragraph.
8. With your teammates, make your presentation to the class.

Evaluation

Discuss your design with your classmates and teacher. What could you do to improve the design?

Name _____ Date _____ Class _____

Tips and Techniques 1

Using Woodworking Tools

Experienced woodworkers often develop techniques that help make their work faster or easier. The following tips will help you get extra benefits from common woodworking tools.

SAFETY FIRST Before doing this activity, make sure you understand how to use all the tools and materials safely.

1. An easy way to set up a miter gauge for a perfect 45-degree angle is to use a framing square. Set the square across the gauge and against the part that pivots, as shown in Fig. 1-1. Align the leg of the miter gauge with the same number on each arm of the square. The miter gauge will then be set at 45 degrees.

2. A framing square can also be used as a guide fence on a drill press. Clamp the square to the table of the drill press so one arm acts as the fence and the other acts as a stop. Fig. 1-2. This technique can help you drill holes accurately in multiple pieces.

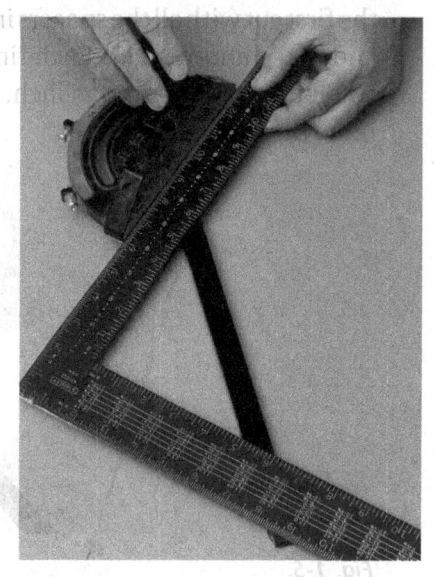

Fig. 1-1.

Fig. 1-2.

3. To check the accuracy of any type of square, align the square with a straightedge and draw a line along the blade on a sheet of paper. Fig. 1-3. Flip the square over and draw another line along the blade in about the same place on the paper. If the lines are parallel, the square is true. If the lines make a V or X, it is not true.

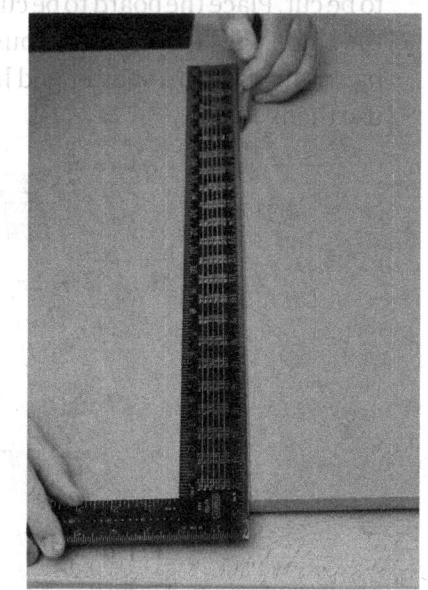

Fig. 1-3.

(Continued on next page)

Name _____ Date _____ Class _____

Tips and Techniques 1 (continued)

4. A combination square can be used to easily draw a straight line down the middle of a board. Align the blade so it is the correct distance from the head of the square. Hold a pencil against the edge of the blade and slide the body of the square along the edge of the board. Fig. 1-4.

5. If you don't have a dado blade, another way to cut a dado is to use a standard ⅛-inch circular-saw blade. First, cut some strips of ⅛-inch hardboard to be used as spacers. Clamp the strips to your table-saw fence. Fig. 1-5. Make the first cut with all the spacers in place. Remove one strip and make another cut. Each time you remove a spacer, you'll take off another ⅛ inch. A ⅜-inch dado requires removing three spacers.

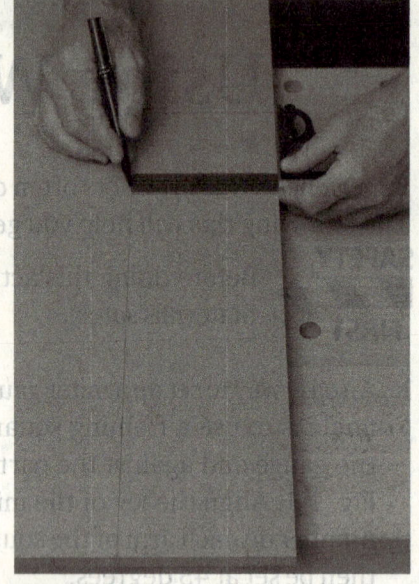

Fig. 1-4.

Fig. 1-5.

6. Ripping thin strips on a table saw can be dangerous. Make it safer by cutting a piece of particleboard or plywood about 6 inches wide and approximately 24 inches long. Attach a handle and a heel to the particleboard. Set the fence to the width of the particleboard plus the width of the strip to be cut. Place the board to be cut up against the particleboard and push it completely through the blade. Fig. 1-6. Be sure to push the piece completely past the blade so as to avoid any kickback. (In Fig. 1-6 a splitter and blade guard are not in place so the cut can be seen. Be sure to use them!)

Fig. 1-6.

(Continued on next page)

166

Name _____ Date _____ Class _____

Tips and Techniques 1 (continued)

7. Hole saws tend to become clogged with sawdust, heat up, and burn the wood. One way to avoid this is to cut a relief hole for the sawdust. First, use the hole saw to lightly score the surface of the wood. Fig. 1-7. Then drill a ¾-inch clearance hole through the workpiece just inside the scored circle. Fig. 1-8. When you resume cutting the large hole, the clearance hole will allow the sawdust to escape.

Fig. 1-7.

Fig. 1-8.

(Continued on next page)

167

Name _____ Date _____ Class _____

Tips and Techniques 2

Making Accurate Measurements

Making accurate measurements is essential to the success of any project. Follow these tips for improved ease and accuracy in measuring.

SAFETY FIRST Before doing this activity, make sure you understand how to use all the tools and materials safely.

1. Don't expect to get accurate measurements from a tape measure that is worn, bent, or marred. Fig. 2-1. If a tape measure is worn out, get a new one and avoid problems.
2. When marking a measurement, make your marks easy to see. Use a sharp pencil or a marking knife. A mechanical pencil with a 0.5-mm hard lead is better than a No.-2 pencil or flat carpenter's pencil. A marking knife creates a precise line. Fig. 2-2.

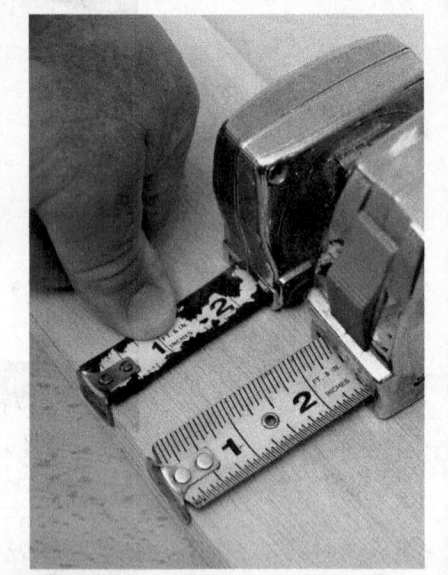

Fig. 2-1.

Fig. 2-2.

3. Mathematical errors are a common cause of inaccurate cuts. A calculator can be a great help, especially one that can work with fractions. Fig. 2-3.

Fig. 2-3.

(Continued on next page)

168

Name _____ Date _____ Class _____

Tips and Techniques 2 *(continued)*

4. It is often safer to measure from a mark on a measuring tape, such as the 10-inch mark, than from the hook at the end of the tape. Fig. 2-4. However, be sure to remember to subtract the starting distance from the overall measurement.

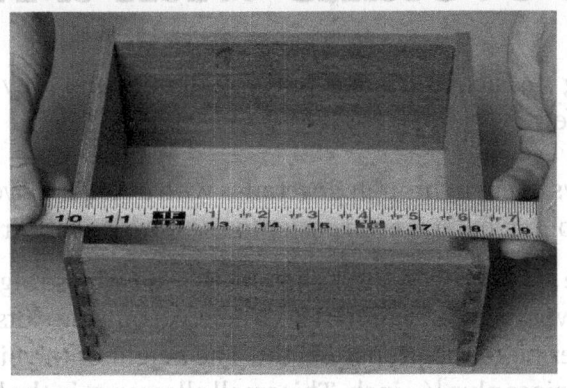

Fig. 2-4.

5. Inside measurements made with a tape measure are much more accurate if you measure twice. Measure an easy-to-read distance out from one corner and make a mark. Next, measure out from the corner opposite the mark. Then add the two measurements together. Hint: Make your first measurement exactly on an inch or foot mark. Then it will be easier to add the other measurement to it.

6. Don't make a measurement if you don't have to, because every measurement or calculation can contain an error. Instead, if possible, measure using the piece of wood that needs to be cut. Fig. 2-5.

Fig. 2-5.

169

Name _____ Date _____ Class _____

Tips and Techniques 3

Cutting Dovetails with a Keller Jig

SAFETY FIRST Before doing this activity, make sure you understand how to use all the tools and materials safely.

One of the simplest ways to cut a through dovetail is with a Keller Dovetail Jig. It is easy to use the jig with a hand-held router, but it is even easier to use it with a table router.

1. Clamp the workpiece to the side of the Keller Jig that has the straight slots. Position the jig template so that the workpiece is centered between two outside slots.
2. Set the dovetail router bit to the proper height. The height of the bit should be equal to the thickness of the workpiece plus $1/32$ inch. This small allowance in the height will be sanded off later when the dovetail is assembled. The bushing of the dovetail router bit will fit perfectly into the slot. Rout the workpieces as shown in Fig. 3-1.

Fig. 3-1.

3. After all the tails have been cut, prepare to rout the pins. Use the tails to lay out the pins on the end of the mating piece. Fig. 3-2. A sharp pencil or bench knife is perfect for this.

Fig. 3-2.

(Continued on next page)

170

Name _____ Date _____ Class _____

Tips and Techniques 3 (continued)

4. Place the flared side of the jig template over the end of the workpiece where the pins will be cut. Align the flared part with the marks you made in Step 3. It should match up perfectly.
5. Place the straight bit in the router and set it to the same height used for the dovetail bit. Rout the pins. Fig. 3-3. It will take two passes to cut the full width of the flared slots.

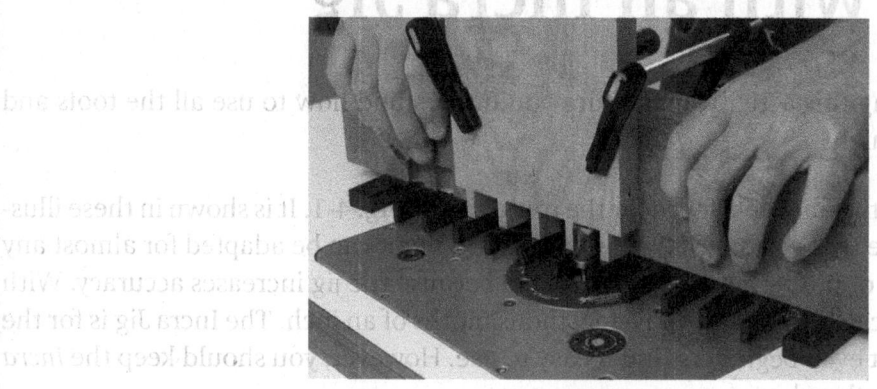

Fig. 3-3.

6. Using a little firm pressure, slide the two pieces together. Fig. 3-4.

Fig. 3-4.

(Continued on next page)

171

Name _____ Date _____ Class _____

Tips and Techniques 4

Making Finger-Lap Joints with an Incra Jig

SAFETY FIRST — Before doing this activity, make sure you understand how to use all the tools and materials safely.

The Incra Jig is a versatile, movable fence for the router table. Fig. 4-1. It is shown in these illustrations on a router table that was specially designed for it, but it can be adapted for almost any router table. Finger-lap, or box, joints are easy to make because the jig increases accuracy. With templates in place, cuts can be made within a few thousandths of an inch. The Incra Jig is for the serious woodworker, but even beginners find it easy to use. However, you should keep the *Incra Master Reference Guide* on hand.

The finger-lap joint is strong and attractive. Its strength comes from the increased area available for gluing. It is used in making drawers and boxes.

1. To make a finger-lap joint with the Incra Jig, select the jig template, the corresponding router bit, and the stock you will be using. Also, cut a piece of ¾-inch stock to the same width as the box sides for use as a backer board. Fig. 4-2.

Fig. 4-1.

Fig. 4-2.

2. Using the *Incra Master Reference Guide*, decide which template to use by placing one of the sides of the box on each template to see which one fits best. Fig. 4-3. Note the suggestions for router bit, stock thickness, and other information along the bottom of the template.
3. Install the appropriate router bit in your router table and set the depth of cut to just a little more than the thickness of the stock. Fig. 4-4.

(Continued on next page)

Name _____ Date _____ Class _____

Tips and Techniques 4 (continued)

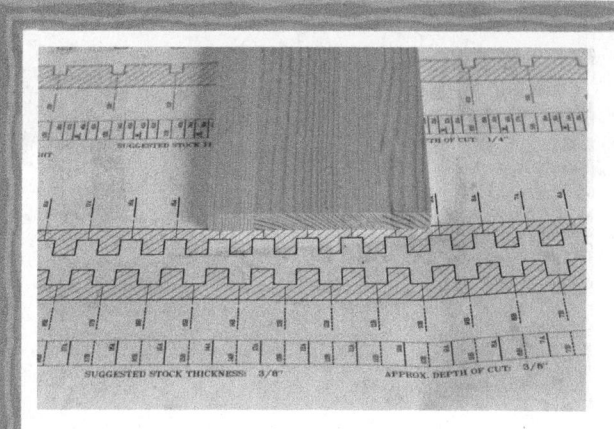

Fig. 4-3.

Fig. 4-4.

4. The ¾-inch backer board will prevent tearout as the bit exits the cut. Be sure the backer board is the same width as the boards you will be using to make the box. Center the router bit by first making a mark at the center of the width of the backer board. Align the mark as closely as possible with the center of the router bit. Set the fence along the edge of the board.
5. Using the right-angle jig, rout a slot in the backer board. Then turn the board around and rout a second slot in the board close to the first slot. This will widen the cut. Fig. 4-5. Adjust the fence until the bit is exactly in the center of the slot.
6. With the jig still locked at the center of the board, push the template strip into a slot in the carriage and slide the recommended "center cut" point under the hairline. Fig. 4-6.

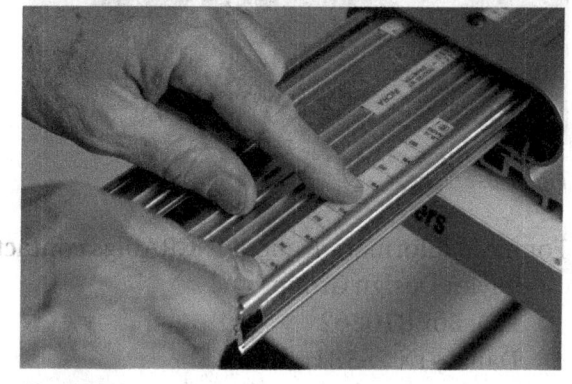

Fig. 4-5.

Fig. 4-6.

7. Clamp two scrap pieces of wood and a backer board to the right-angle fixture and make two or three practice cuts. Separate the two test pieces and check the fit. Fig. 4-7.
8. Clamp the two sides of the box to the right-angle jig and make the cuts at the *A* marks on the template. Fig. 4-8. (Naturally, the only cuts that need to be made are those that fit within the width of the boards.) After completing the cuts, flip the boards end for end and repeat the same *A* cuts on the other end.

(Continued on next page)

173

Tips and Techniques 4 (continued)

Fig. 4-7.

Fig. 4-8.

9. Clamp the two end boards of the box to the right-angle fixture and make the B cuts at each end. Fig. 4-9.
10. Glue the sides and ends together for perfect finger-lap joints. Fig. 4-10.

Fig. 4-9.

Fig. 4-10.

For more information about INCRA Jigs contact:
Taylor Design Group
P.O. Box 810262
Dallas, TX 75381
www.incra.com
or
Woodpecker, Inc.
11050 Industrial First Avenue
North Royalton, OH 44133
www.woodpeck.com

174

Name _____ Date _____ Class _____

Tips and Techniques 5

Making Half-Blind Dovetail Joints with an Incra Jig

SAFETY FIRST Before doing this activity, make sure you understand how to use all of the tools and materials safely.

The half-blind dovetail joint is easy to make with the Incra Jig. This joint is strong and is often used in fine furniture construction, especially for drawers. It is easy to make because the jig increases accuracy. With templates in place, cuts can be made within a few thousandths of an inch. The Incra Jig is for the serious woodworker, but even beginners find it easy to use. However, you should keep the *Incra Master Reference Guide* on hand.

1. Begin by selecting the template, corresponding bit, and stock width for the box or drawer. Also, cut a piece of ¾-inch thick stock to the same width as the box sides. This piece will be used as a backer board to prevent tearout. Fig. 5-1. Note that in the *Incra Master Reference Guide*, stock thickness for the dovetail is suggested under each template.
2. Install the appropriate router bit in the router and set the depth of cut to just a little more than the thickness of the stock. Fig. 5-2.

Fig. 5-1.

Fig. 5-2.

3. Be sure the backer board is the same width as the boards you will be using for the box. Center the router bit by first making a mark at the center of the width of the backer board. Align the mark as closely as possible with the center of the router bit. Set the fence along the edge of the board.
4. Using the right-angle jig, rout a tail cut in the backer board. Then turn the board around and rout another tail cut in the board very close to the first cut. This will widen the cut. Adjust the fence until the bit is positioned exactly in the center of the cut. Fig. 5-3.

(Continued on next page)

175

Name _____ Date _____ Class _____

Tips and Techniques 5 (continued)

5. With the jig still locked at the center of the board, push the template strip into a slot in the carriage and slide the recommended "center cut" point under the hairline. Fig. 5-4.

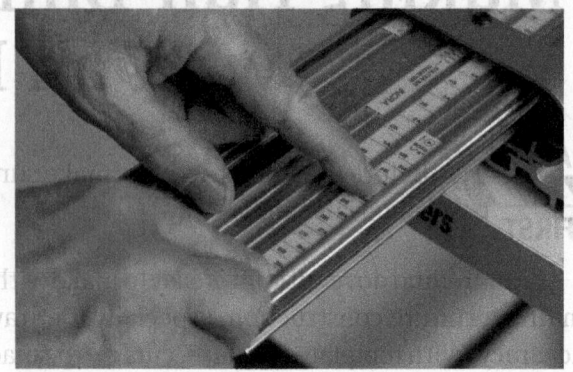

Fig. 5-3.

Fig. 5-4.

6. To cut the tails, cut a dovetail-shaped rabbet on both ends of two test pieces that are the same thickness as those in the finished box. Fig. 5-5. The dimension for the rabbet cut is listed under the diagram of the template in the Reference Guide. Do not attempt to cut the rabbet in one pass. Make two or three light cuts so as not to cut too deep.
7. Cut tails in the test pieces. Fig. 5-6. Fit the two test pieces together to check for depth of cut. Fig. 5-7.

Fig. 5-5.

Fig. 5-6.

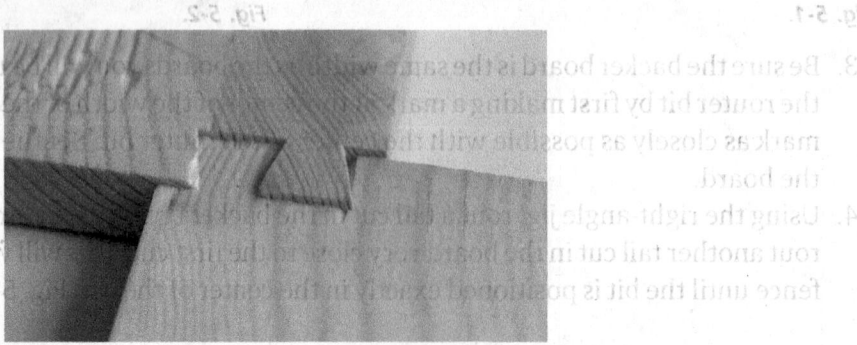

Fig. 5-7.

(Continued on next page)

176

Name _____ Date _____ Class _____

Tips and Techniques 5 (continued)

8. Make rabbet cuts on the ends of the two sides of the box. Clamp the sides and the backer board to the right-angle fixture. Fig. 5-8. The rabbet cuts should be on the outside. Then make the tail cuts. Flip the boards end for end and repeat the cuts.

9. Set the fence for the first pin cut on the template that will place the bit in front of the fence. Set an Incra stop as close as possible to the dovetail router bit without touching it. The stop is used to limit the length of the pin cuts. Fig. 5-9.

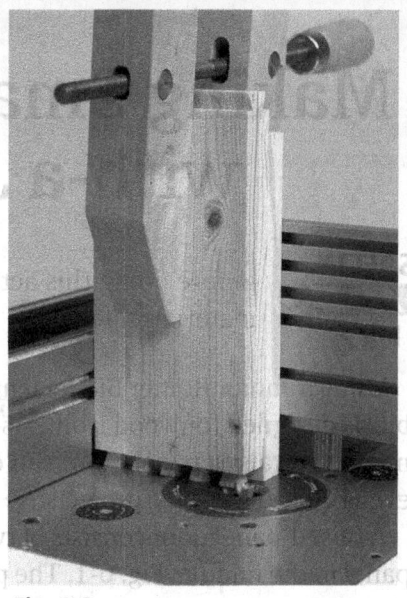

Fig. 5-8.

Fig. 5-9.

10. Place one of the end boards face down on the router table and make the series of pin cuts. Be sure to use a rubber-soled pushblock, as shown in Fig. 5-10. Make the cuts on only one end of the board and check the fit with the tail cuts. Fig. 5-11. If the tails don't fit all the way into the pins, simply lengthen the pin cuts by moving the stop away from the bit and re-cut.

11. Apply glue and assemble the box.

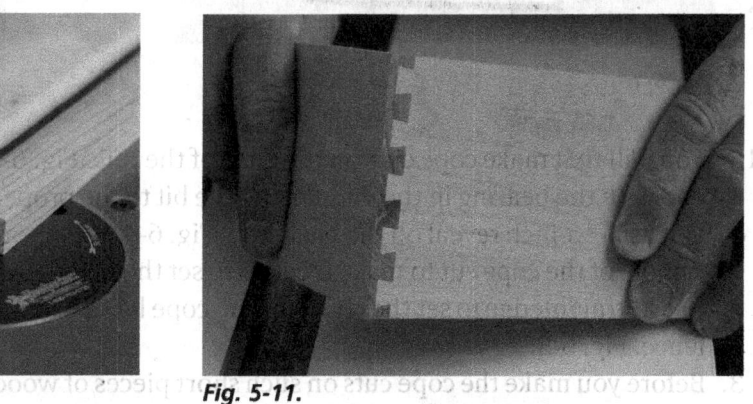

Fig. 5-10.

Fig. 5-11.

For more information about INCRA Jigs contact:

Taylor Design Group
P.O. Box 810262
Dallas, TX 75381
www.incra.com

or

Woodpecker, Inc.
11050 Industrial First Avenue
North Royalton, OH 44133
www.woodpeck.com

177

Name _____ Date _____ Class _____

Tips and Techniques 6

Making Small Raised-Panel Doors with a Junior Router Set

SAFETY FIRST Before doing this activity, make sure you understand how to use all the tools and materials safely.

Using a Junior Router Bit Set is an inexpensive way to make small raised-panel doors. The set of bits is designed for wood from $7/16$- to $11/16$-inch thick. After you have learned how to make these miniature raised panels, you can easily move on to full-size doors. The small raised panels can even be used for some projects.

A raised-panel door consists of two stiles, two rails, and the panel in the middle. Fig. 6-1. The panel slips into the stiles. Fig. 6-2.

Fig. 6-1.

Fig. 6-2.

1. You will first make cope cuts on the ends of the rails. Fig. 6-3. Use the cope-cutting bit, the one with the bearing in the middle. Set the bit to the proper height. The cope should have at least a $1/8$-inch reveal on the back side. Fig. 6-4. After you set the height, make and keep a sample of the cope cut to make it easier to set the height for your next project.
2. Use a straightedge to set the fence for the cope bit. Align the bearing on the cope bit with the fence. Fig. 6-5.
3. Before you make the cope cuts on such short pieces of wood, take several safety precautions. First, be sure the area around the router bit is small enough so the wood will not fall into it. Be sure the fence is tight around the bit. Attach a plastic push pad to a pushblock and use it to push the small pieces past the bit. Fig. 6-6. Also, cut a block from a 2×4 and notch it like the one shown in Fig. 6-6. This block can be used to hold the work against the fence as the pushblock pushes the stock through the bit. Fig. 6-7.

(Continued on next page)

178

Name _____ Date _____ Class _____

Tips and Techniques 6 (continued)

Fig. 6-3.

Fig. 6-4.

Fig. 6-5.

Fig. 6-6.

4. Make all panels with the good side of the stock face down on the table. The part you will see as you work is the inside of the raised panel.
5. Make the pattern cuts. The pattern cuts go on the inside of the stiles and rails and hold the raised panel in place. See Figs. 6-2 and 6-3 again. Set the height of the pattern bit so that the tongue of the cope cut lines up with the top cutter of the pattern bit. Fig. 6-8. Align the fence with the bearing, as you did for the cope bit in Step 2.

Fig. 6-7.

Fig. 6-8.

(Continued on next page)

179

Name _____ Date _____ Class _____

Tips and Techniques 6 (continued)

6. To better grip the workpiece as you push it through the router, first run the pushblock through the pattern bit to make a pattern cut on its end. Fig. 6-9. The cope cut will now fit right into the pattern cut, and you will have better control as you push it past the bit. Fig. 6-10.

Fig. 6-9.

Fig. 6-10.

7. You have completed the rails and the stiles. Fig. 6-11. To make the raised panel, set the raised-panel cutter so the bearing lines up with the groove that was cut with the pattern bit. Fig. 6-12.
8. Clamp an auxiliary fence onto the regular fence. Cut an opening in the auxiliary fence with the router bit so there is zero clearance all around the bit. Fig. 6-13. This will keep the small workpieces from falling into any gap around the bit.
9. Because the router bit is so big, slow the router speed down as you make the cut. Use the pushblock and side block.
10. Slip the panel into the stiles and rails. You are now ready to make a project like those shown in Fig. 6-14.

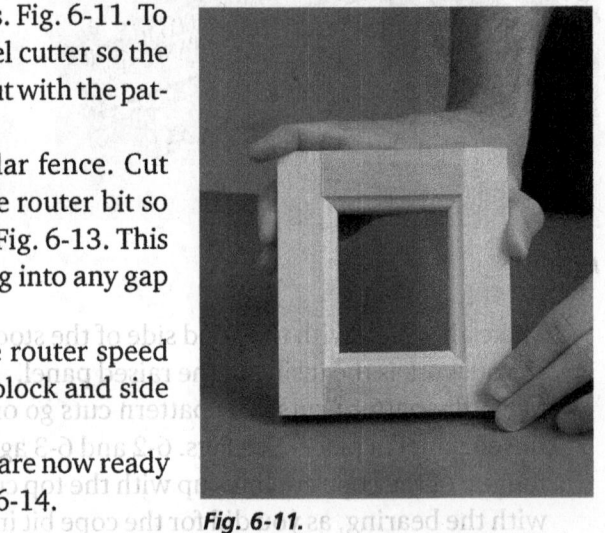

Fig. 6-11.

Fig. 6-12.

Fig. 6-13.

(Continued on next page)

180

Name _____ Date _____ Class _____

Tips and Techniques 6 (continued)

Fig. 6-14.

For more information about the Junior Raised-Panel Set, including plans and DVDs, contact:
Sommerfeld's Tools for Wood
1408 Celebrity Rd. HWY 3 West
Remsen, IA 51050
www.sommerfeldtools.com

Name _____ Date _____ Class _____

Tips and Techniques 7

Using Clamps

SAFETY FIRST Before doing this activity, make sure you understand how to use all of the tools and materials safely.

Following is a brief overview of several different kinds of clamps. As you read, you may think of additional applications for each kind.

1. The bar clamp is one of the least expensive and most versatile. Fig. 7-1. You can make it almost any length you desire just by adding couplings.
2. The jaws of the K-Body Clamp stay parallel to each other. Fig. 7-2. This is most useful when you are gluing a panel. K-Body Clamps come in several lengths.

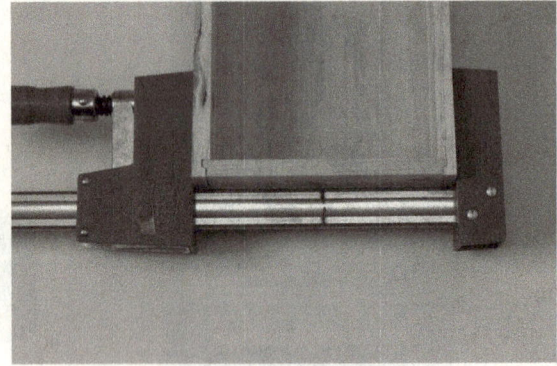

Fig. 7-1.

Fig. 7-2.

3. Angle clamps are good for clamping a mitered corner or a butt joint. Fig. 7-3. Nails or screws can be applied while the workpiece is still in the clamp. This makes the joint even stronger. Fig. 7-4.

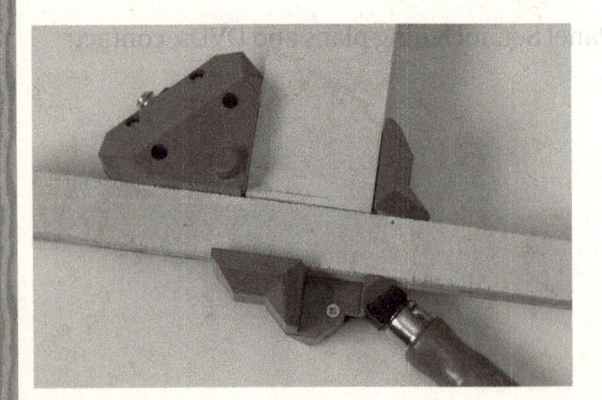

Fig. 7-3.

Fig. 7-4.

(Continued on next page)

182

Name _____ Date _____ Class _____

Tips and Techniques 7 (continued)

4. The Klik Clamp has a ratchet for quick clamping and release. Fig. 7-5. Because it is made from magnesium, it is very lightweight and is useful for lightweight workpieces.
5. Small bar clamps are useful for clamping lightweight workpieces like the lamination shown in Fig. 7-6. They come in many different lengths.

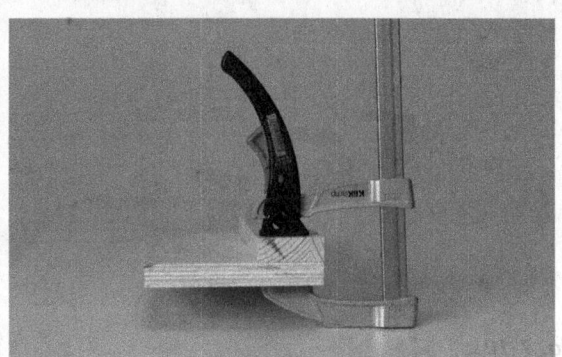

Fig. 7-5.

Fig. 7-6.

6. The band clamp makes it easy to clamp workpieces containing almost any angle. Fig. 7-7. Small corner attachments can be added to the band to protect the corners.

Fig. 7-7.

7. The Super Grip Clamp adjusts quickly. Fig. 7-8. It is usually used for holding pieces together lightly rather than applying a lot of pressure.

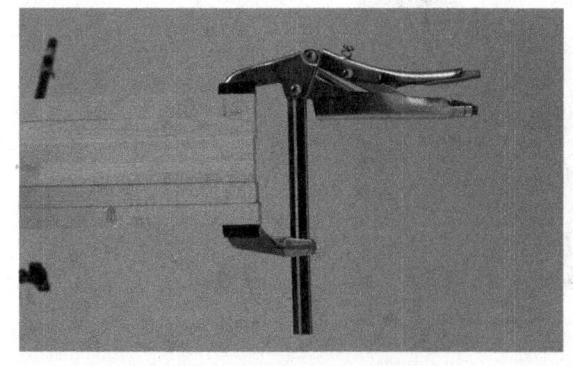

Fig. 7-8.

(Continued on next page)

Name _____ Date _____ Class _____

Tips and Techniques 7 (continued)

8. An edge clamp is used when applying an edge to a tabletop or countertop. Fig. 7-9. A special attachment for edge clamping can be used along with a bar clamp. Fig. 7-10.

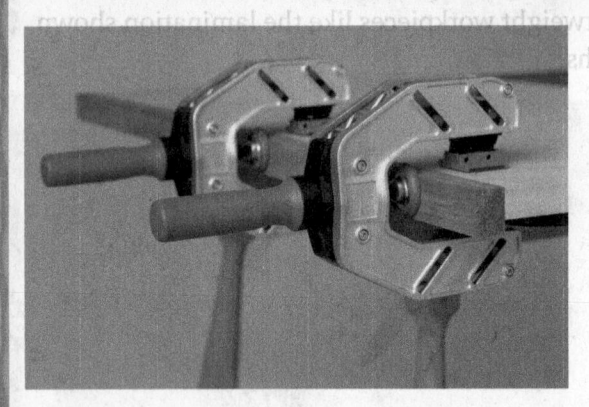

Fig. 7-9.

Fig. 7-10.

9. The VarioPivot Clamp can be used along with a K-Body Clamp to hold workpieces at almost any angle. Fig. 7-11.

10. Hand-screw clamps are versatile and can be used in many different ways. Fig. 7-12. They can be adjusted so their jaws are not parallel.

Fig. 7-12.

Fig. 7-11.

(Continued on next page)

184

Name _____ Date _____ Class _____

Tips and Techniques 8

Making Pocket-Hole Joints

SAFETY FIRST Before doing this activity, make sure you understand how to use all of the tools and materials safely.

Adding pocket-hole screws to strengthen joints has become more popular since inexpensive equipment became available for small shops and home use. Fig. 8-1. Pocket-hole joinery is much faster than other wood-joining techniques because the self-tapping screws that are used eliminate the need to pre-drill the mating workpiece. Fig. 8-2. This prevents any misalignment problems-as well. The self-tapping screw also acts as a clamp while the glue used in the joint can cure. The process is speeded up because you do not have to wait for the glue to cure before removing any clamps.

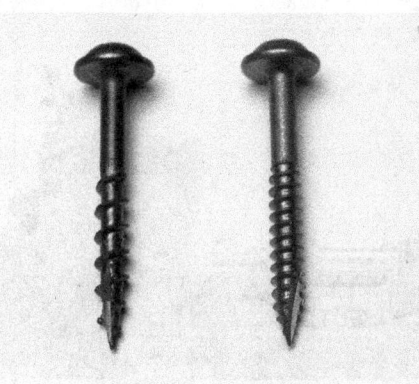

Fig. 8-1.

Fig. 8-2.

Following are a few basic techniques that will help you make pocket-hole joints.

1. Drilling pocket holes can be done in several ways. Expensive cabinet shops may use a dedicated machine just for drilling holes. Small shops or hobbyists often use a drill and a K2000 Kreg Jig. Fig. 8-3. You simply clamp the workpiece into the jig and drill the hole.

Fig. 8-3.

(Continued on next page)

Tips and Techniques 8 (continued)

2. The depth of the hole depends on the thickness of the wood. Fig. 8-4. In Fig. 8-5, you can see how the jig is set to different heights and how inserts are placed in the jig for various thicknesses of wood.

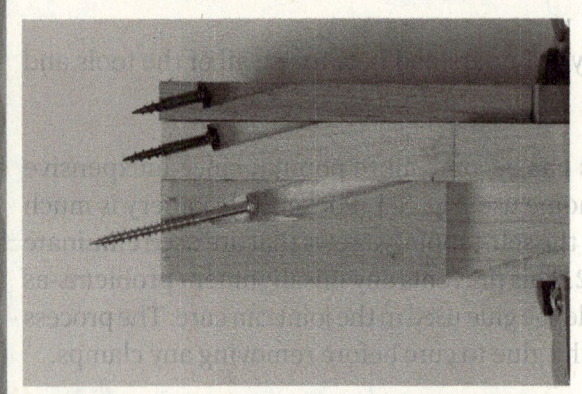

Fig. 8-4.

Fig. 8-5a.

Fig. 8-5b.

Fig. 8-5c.

3. The stop collar on the drill must also be adjusted for different thicknesses. It can be set on the arm of the K2000 or by actually measuring the distance from the drill flute (not the tip) to the collar. Fig. 8-6.

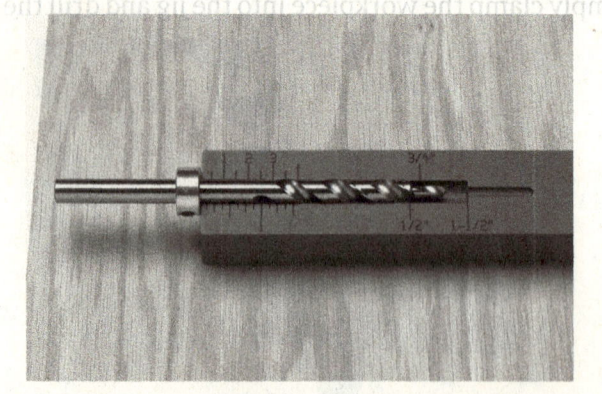

Fig. 8-6.

(Continued on next page)

Name _____ Date _____ Class _____

Tips and Techniques 8 (continued)

4. On the top of the jig are guide holes for drilling holes spaced differently. Fig. 8-7. On a typical stile or rail you would normally use holes *A* and *B*.
5. Another way to drill pocket holes is with a Rocket Jig. See Fig. 8-1 again. To use this jig, simply clamp it to the workpiece and drill the holes. Fig. 8-8.
6. For special repair jobs, you can use the even smaller Mini Jig, which can be clamped to a finished workpiece. In Fig. 8-9, it has been clamped to a chair to repair and strengthen the legs.
7. To insert the screws in the pocket holes, first clamp the mating pieces together, being sure the face side is perfectly flat. The clamp shown in Fig. 8-10 is perfect for this, or you can make a special jig like the one shown in Fig. 8-11.

Fig. 8-7.

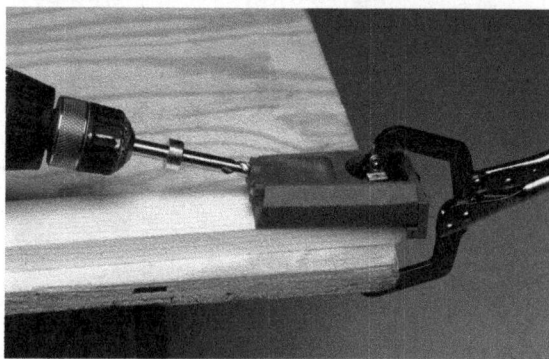

Fig. 8-8.

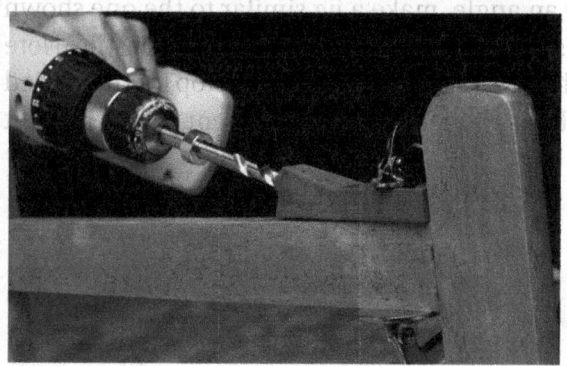

Fig. 8-9.

Fig. 8-10.

Fig. 8-11.

(Continued on next page)

187

Name _____ Date _____ Class _____

Tips and Techniques 8 (continued)

8. Special plugs can be used to fill pocket holes after the screws are in place. Fig 8-12. Several different types of wood are available as well as plastic for MDF and melamine.
9. When you are building cabinets, one cabinet may sometimes be installed at an angle to the adjoining cabinet. An easy way to join the two stiles is to cut the angle on only one stile. For example, if the cabinets are joined at a 45-degree angle, you would cut only one stile at 45-degrees. This is much easier than trying to cut each stile at 22½ degrees and then miter them together. When you are finished, there will be a small overhang that must be sanded or cut off. Fig. 8-13.

Fig. 8-12.

Fig. 8-13.

10. To screw together the two stiles of cabinets at an angle, make a jig similar to the one shown in Fig. 8-14. The angle on the piece at the left is the same angle that was cut on the stile. Note that there is also a shim attached to the base. This creates a V that acts like a stop for the angled stile. For ¾-inch stock, this shim is typically about ¼-inch thick. Experiment with it to get it perfect.
11. Slip the angled stile into the groove, as shown in Fig. 8-15, and hold it in place.

Fig. 8-14.

Fig. 8-15.

(Continued on next page)

Tips and Techniques 8 (continued)

12. Place the straight stile with the pocket holes drilled in it into the jig next to the angled workpiece and screw them together. Fig. 8-16.
13. After all the screws have been inserted, use a belt sander to sand off the overhang on the joint. Fig. 8-17.

Fig. 8-16.

Fig. 8-17.

14. Because the overhang has been removed, the two stiles will not appear to be the same width unless they are initially cut to different widths. For example, for the stiles to look as though both are 2 inches wide, the stile with the straight edge should be cut at 1⅝ inches, and the stile with the angled edge should be cut at 2⅜ inches. Fig. 8-18. After the joint is completed, it is very difficult to see the glue line.

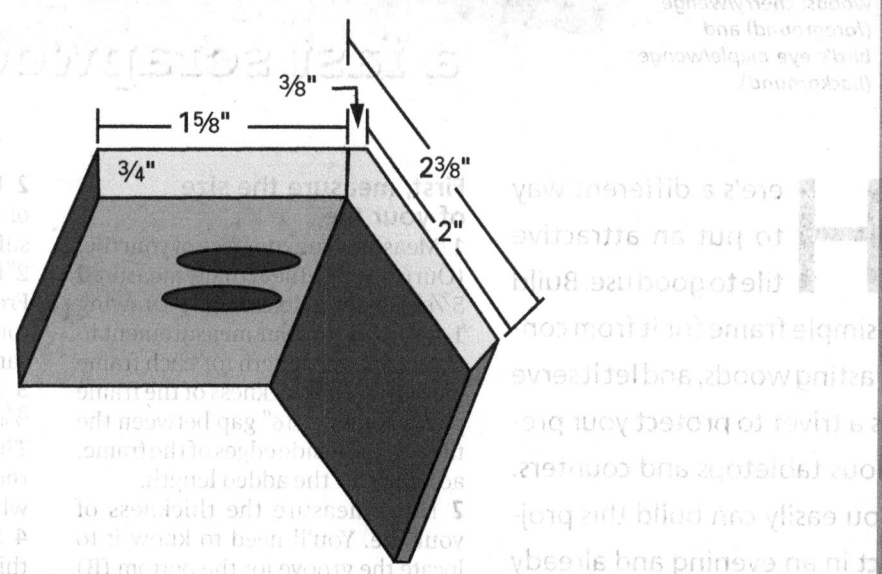

Fig. 8-18.

For information about Kreg pocket-hole screws and jigs contact:
Kreg Tool Company
201 Campus Drive
Huxley, IA 50124
www.kregtool.com

Project 1   189

kitchen trivet
a fast scrapwood project

Attractive tiles and contrasting woods give instant eye appeal to these trivets. The featured woods: cherry/wenge (foreground) and bird's-bird's-eye maple/wenge (background).

Here's a different way to put an attractive tile to good use. Build a simple frame for it from contrasting woods, and let it serve as a trivet to protect your precious tabletops and counters. You easily can build this project in an evening and already may have the items needed on hand: a 6" or 8" square tile of your choice, a couple strips of wood, and a small piece of 1/8" hardboard.

First, measure the size of your tile

1 Measure the actual size of your tile. (Our 6"-square tile actually measured 5 7/8" square.) Referring to Drawing 1, add 1 5/8" to your measurement to determine the length for each frame piece (A). The thickness of the frame pieces, plus a 1/16" gap between the tile and the inside edges of the frame, accounts for the added length.

2 Now, measure the thickness of your tile. You'll need to know it to locate the groove for the bottom (B) later.

Make the parts, assemble, and finish

1 Select two 3/4"-thick contrasting woods to form a blank for the frame pieces (A).

2 From the stock selected for the top of the frame, cut a 1"-wide blank of sufficient length (including at least 2" for waste) to form the frame pieces. From the stock selected for the bottom part, cut a 1/4"-wide blank of the same length.

3 Edge-join the blanks to form a 3/4×1 1/4" blank and sand smooth. Then, rout a 1/2" round-over along the top outside edge of the blank, where shown on Drawing 1.

4 Subtract 1/8" from the measured thickness of your tile. The result is the distance from the top of the groove that will receive the bottom (B) to the top of the frame, where shown on Drawing 1a. This will set the tile's top 1/8" above the frame, where shown. Cut a 1/8" groove 1/4" deep along the inside face of the blank.

190 Project 1 Article adapted from WOOD magazine, issue 146, December 2002

1 EXPLODED VIEW

- 6" or 8" square tile
- 1" hole, centered
- ⅛" groove ¼" deep
- ½" round-overs
- B
- Width of tile + ⅝"
- A
- A
- Width of tile + 1⅝"

1a FULL-SIZE SECTION VIEW

- Thickness of tile − ⅛"
- Tile
- ⅛"
- 1/16"
- ⅛"
- 1"
- A
- ¼"
- B
- ¾"
- R = ⅝"
- ¼"
- 1⅝"

Bandsaw to the layout line to form the cutout in the frame pieces. Sand to final shape using a drum sander.

A

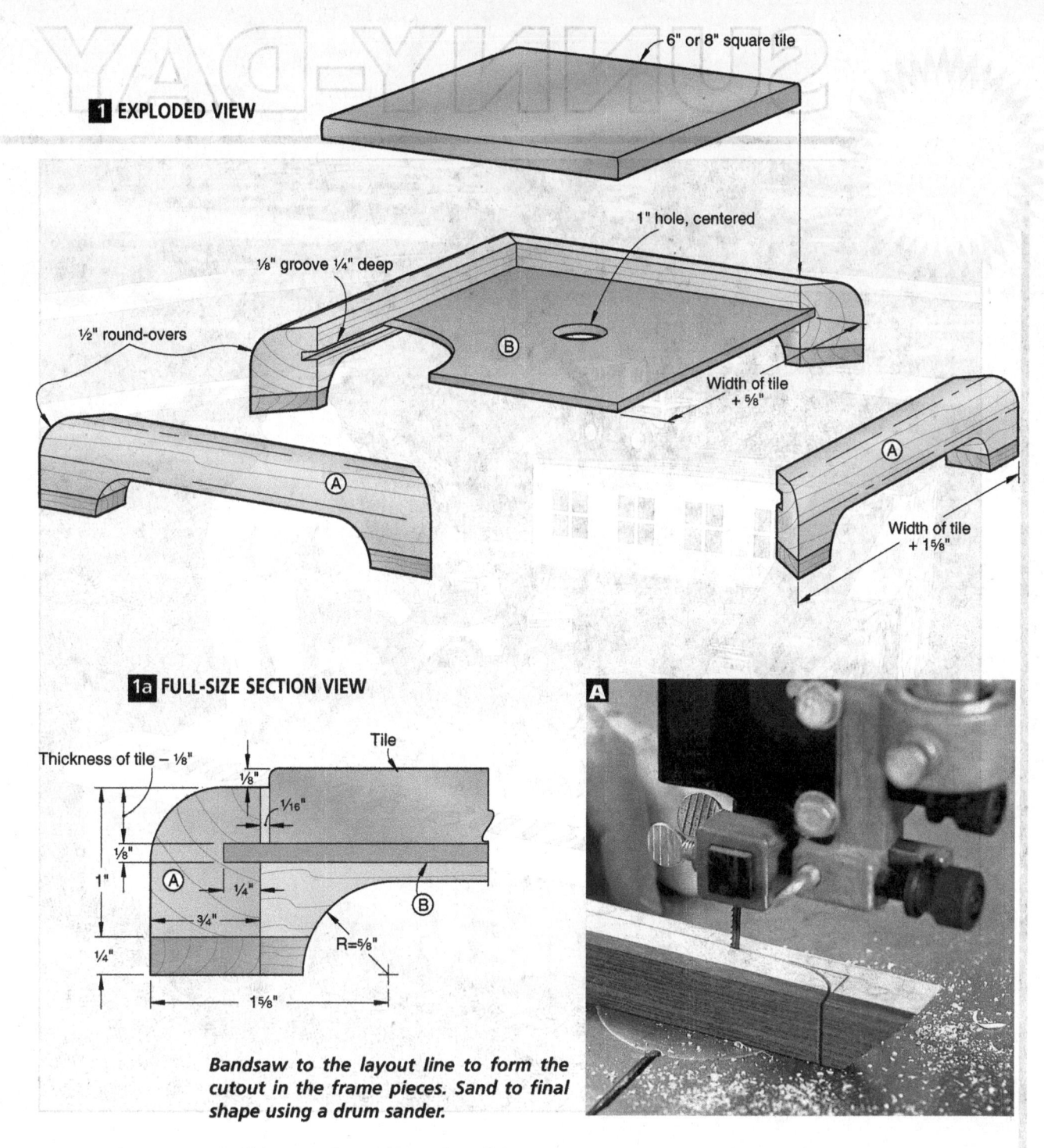

5 Miter-cut the blank to form the frame pieces (A) to the length shown on Drawing 1. Use an auxiliary extension on your miter gauge as a backer board to prevent chip-out, and use a stopblock to ensure you cut all of the pieces to the same length.

6 Mark the cutout at the bottom of each frame piece by marking a ⅝" radius at each end, where dimensioned, and drawing a line between them. Referring to Photo A, bandsaw the cutouts to shape, and sand smooth.

7 From ⅛" hardboard, cut the bottom (B) so it's ⅝" larger than your tile's actual size. Draw diagonals to find the bottom's center, then drill a 1" hole through the bottom, where shown. This hole lets you remove the tile for cleaning or replacement.

8 Dry-assemble the frame and the bottom, and verify a good fit. Then, apply glue in the grooves in the frame pieces and to their mitered ends. Assemble and secure with a band clamp, checking for tight miter joints.

9 Finish-sand the frame to 220 grit, and remove the dust. Apply a clear finish of your choice. (We brushed-on two coats of satin polyurethane, sanding to 320 grit between coats.) Now, place your tile in the frame, and put your trivet into service.

SUNNY-DAY

Hey parents and grandparents, are you looking for the perfect way to keep your kids off the streets and out of trouble? We've got the perfect solution. Our sandbox/toy storage center will keep your kids, and probably many of your neighbors' kids, happily occupied for many fun-filled hours all in the safety of your backyard. And as you can see by looking at the drawings here, this kid-pleasing project is a breeze to build. We completed ours (not including the painting) in just a day. You can, too!

Start with the toy box garage

1 From ¾" exterior plywood, cut the front and back (A), ends (B), and bottom (C) to the sizes listed in the Bill of Materials.

2 Cut a ¾" groove ¼" deep ¾" from the bottom edge of the front, back, and end plywood panels (A, B). Cut a ¾" rabbet ½" deep across both ends of the front and back panels.

192 **Project 2**

Article adapted from WOOD magazine, issue 71, August 1994

SANDBOX

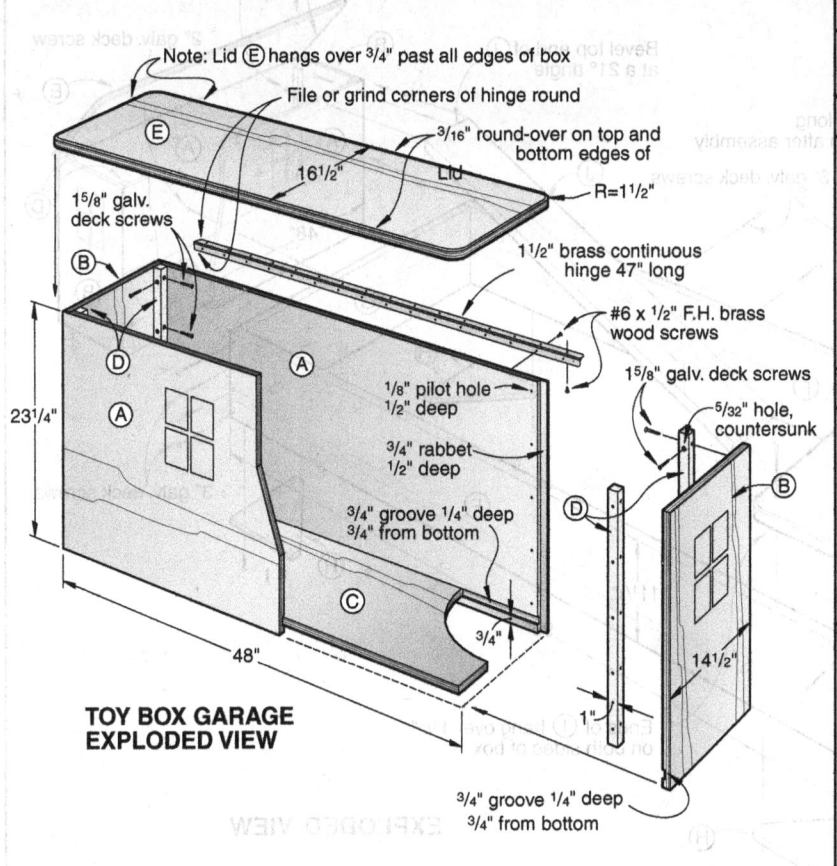

TOY BOX GARAGE EXPLODED VIEW

Note: Lid E hangs over 3/4" past all edges of box
File or grind corners of hinge round
3/16" round-over on top and bottom edges of Lid
16 1/2"
R=1 1/2"
1 5/8" galv. deck screws
1 1/2" brass continuous hinge 47" long
#6 x 1/2" F.H. brass wood screws
1/8" pilot hole 1/2" deep
1 5/8" galv. deck screws
5/32" hole, countersunk
3/4" rabbet 1/2" deep
23 1/4"
3/4" groove 1/4" deep 3/4" from bottom
48"
3/4"
14 1/2"
1"
3/4" groove 1/4" deep 3/4" from bottom

Bill of Materials

Part	T	W	L	Matl.	Qty.
TOY BOX GARAGE					
A front & back	3/4"	23 1/4"	48"	XP	2
B ends	3/4"	14 1/2"	23 1/4"	XP	2
C bottom	3/4"	14"	47"	XP	1
D cleats	3/4"	1"	21 3/4"	C	4
E lid	3/4"	16 1/2"	49 1/2"	XP	1
SANDBOX					
F sides	1 1/2"	11 1/4"	96"	C	2
G ends	1 1/2"	11 1/4"	45"	C	2
H gussets	3/4"	12"	12"	XP	4
I seat	1 1/2"	11 1/4"	51"	C	1
J ramp	1 1/2"	9 1/4"	60"	C	1

Material Key: XP–exterior plywood
C–choice (fir, spruce, pine, redwood)

Supplies: 1 5/8", 2", 2 1/2", and 3" galvanized deck screws, multi-purpose adhesive, 1 1/2 × 47" brass continuous (piano) hinge, #6 × 1/2" flathead brass wood screws, enamel paints

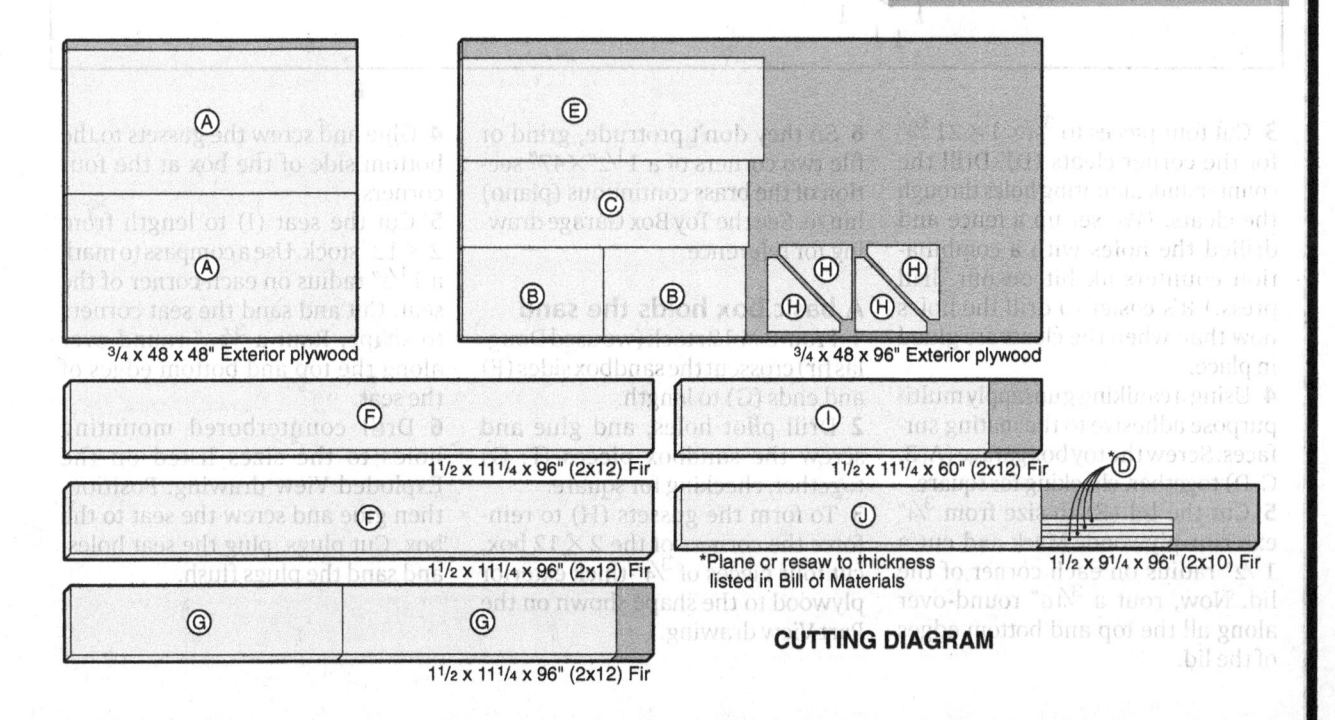

3/4 x 48 x 48" Exterior plywood

3/4 x 48 x 96" Exterior plywood

1 1/2 x 11 1/4 x 96" (2x12) Fir

1 1/2 x 11 1/4 x 96" (2x12) Fir

1 1/2 x 11 1/4 x 96" (2x12) Fir

1 1/2 x 11 1/4 x 60" (2x12) Fir

*Plane or resaw to thickness listed in Bill of Materials

1 1/2 x 9 1/4 x 96" (2x10) Fir

CUTTING DIAGRAM

Article adapted from WOOD magazine, issue 71, August 1994

SUNNY-DAY SANDBOX

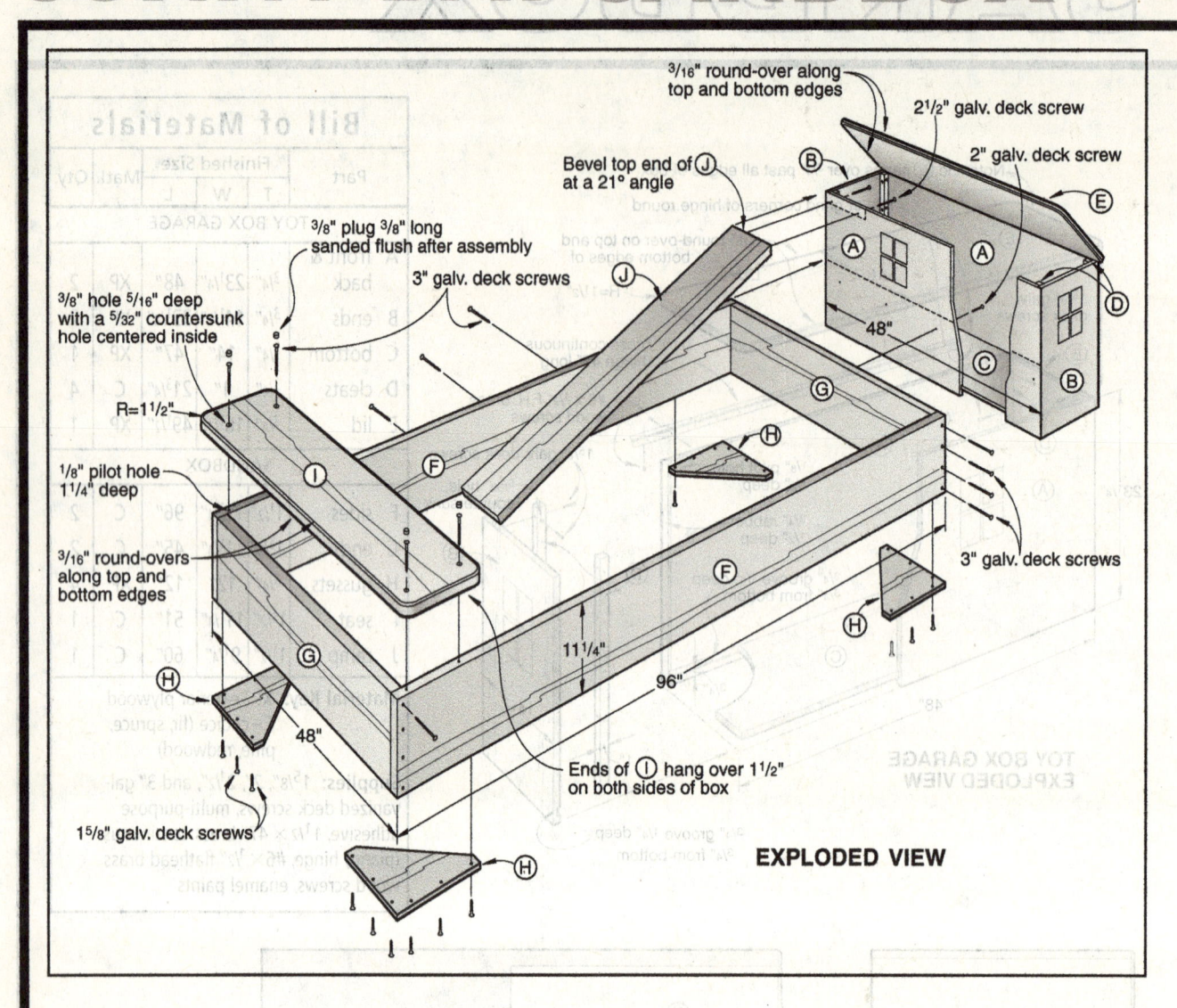

Exploded View labels:
- 3/16" round-over along top and bottom edges
- 2 1/2" galv. deck screw
- 2" galv. deck screw
- Bevel top end of J at a 21° angle
- 3/8" plug 3/8" long sanded flush after assembly
- 3" galv. deck screws
- 3/8" hole 5/16" deep with a 5/32" countersunk hole centered inside
- R = 1 1/2"
- 1/8" pilot hole 1 1/4" deep
- 3/16" round-overs along top and bottom edges
- 3" galv. deck screws
- 11 1/4"
- 96"
- 48"
- Ends of I hang over 1 1/2" on both sides of box
- 1 5/8" galv. deck screws

EXPLODED VIEW

---

3 Cut four pieces to 3/4 × 1 × 21 3/4" for the corner cleats (D). Drill the countersunk mounting holes through the cleats. (We set up a fence and drilled the holes with a combination countersink bit on our drill press.) It's easier to drill the holes now than when the cleats are glued in place.

4 Using a caulking gun, apply multipurpose adhesive to the mating surfaces. Screw the toy box garage (A, B, C, D) together, checking for square.

5 Cut the lid (E) to size from 3/4" exterior plywood. Mark and cut a 1 1/2" radius on each corner of the lid. Now, rout a 3/16" round-over along all the top and bottom edges of the lid.

6 So they don't protrude, grind or file two corners of a 1 1/2" × 47" section of the brass continuous (piano) hinge. See the Toy Box Garage drawing for reference.

A basic box holds the sand

1 From 2 × 12 stock (we used Douglas fir) crosscut the sandbox sides (F) and ends (G) to length.

2 Drill pilot holes, and glue and screw the sandbox pieces (F, G) together, checking for square.

3 To form the gussets (H) to reinforce the corners of the 2 × 12 box, cut four pieces of 3/4"-thick exterior plywood to the shape shown on the Part View drawing.

4 Glue and screw the gussets to the bottom side of the box at the four corners.

5 Cut the seat (I) to length from 2 × 12" stock. Use a compass to mark a 1 1/2" radius on each corner of the seat. Cut and sand the seat corners to shape. Rout a 3/16" round-over along the top and bottom edges of the seat.

6 Drill counterbored mounting holes to the sizes listed on the Exploded View drawing. Position, then glue and screw the seat to the box. Cut plugs, plug the seat holes, and sand the plugs flush.

194 Project 2

Article adapted from WOOD magazine, issue 71, August 1994

SECTION VIEW

Ramp (J) is secured to Side Rail (F) with 3" galv. deck screws

PART VIEW

GUSTSET (H)

5/32" holes, countersunk

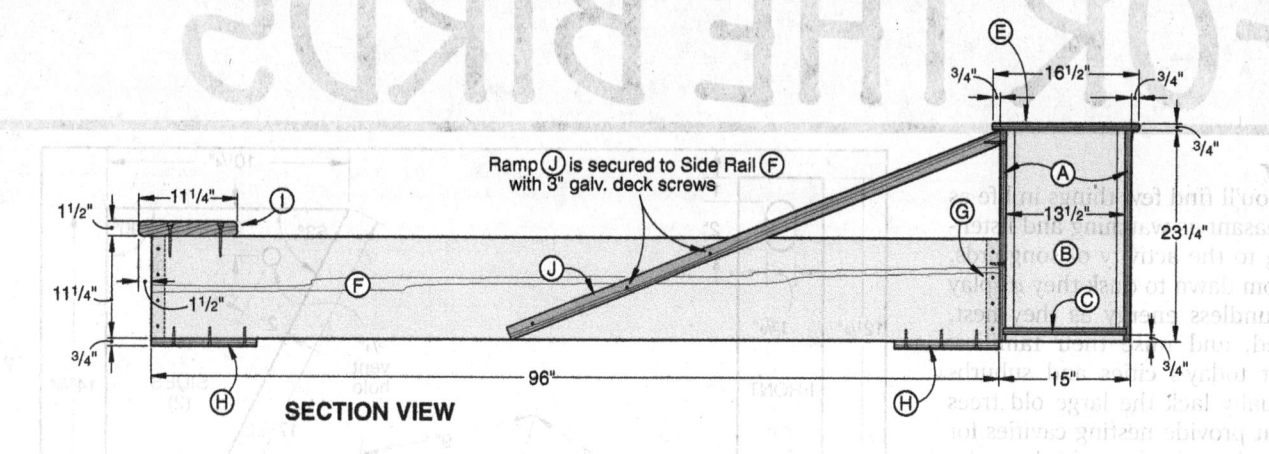

7 Cut the ramp (J) to length from a 2×10, cutting one end at 21° where shown on the Exploded View drawing.

Final touches before ordering the sand

1 Sand the sandbox and toy box. Fill any imperfections with a wood filler. Sand the filled areas.
2 Apply a coat of primer to both assemblies (don't forget to paint the inside of the toy box garage). (See the box *below right* for painting particulars.) Paint the sandbox and toy box.
3 For an added effect, detail-paint the windows on the toy box garage and the logo on the sandbox seat. For the seat logo, we used 5" deck stencil lettering.
4 Drill the screw holes and screw the toy box garage to the sandbox.
5 Position the lid on the toy box, and then clamp the continuous hinge in place. Screw the hinge to the bottom side of the lid and to the back side of the toy box.
6 Position and then screw the painted ramp in place. (We used galvanized screws to fasten the ramp to the garage front and to the side of the sandbox.)

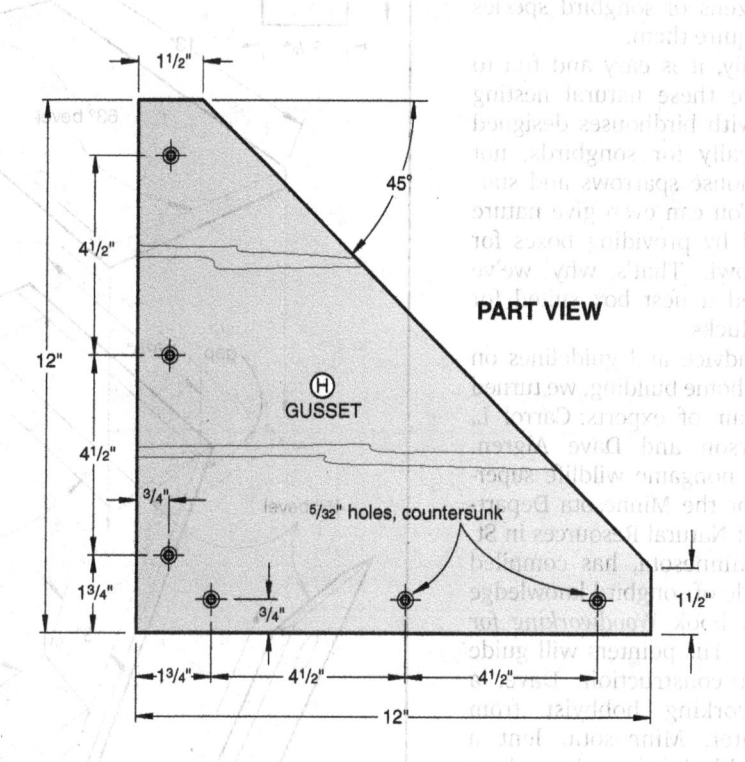

How we painted our sandbox

We first primed the toy box garage and sandbox pieces with one coat of an oil-based primer. We then applied two coats of an industrial oil-based enamel, letting each coat dry at least 24 hours. For the lettering and windows, we used an enamel paint. To paint the windows, we masked the outlines with masking tape and removed the tape within an hour after painting. Left on too long, the tape will leave a sticky residue that's hard to remove.

Article adapted from WOOD magazine, issue 71, August 1994

Project 2

HOUSING FOR THE BIRDS

You'll find few things in life as pleasant as watching and listening to the activity of songbirds. From dawn to dusk they display boundless energy as they nest, feed, and raise their families. But today's cities and suburbs usually lack the large old trees that provide nesting cavities for the dozens of songbird species that require them.

Luckily, it is easy and fun to simulate these natural nesting spots with birdhouses designed specifically for songbirds, not pesty house sparrows and starlings. You can even give nature a hand by providing boxes for waterfowl. That's why we've included a nest box suited for wood ducks.

For advice and guidelines on proper home building, we turned to a pair of experts: Carrol L. Henderson and Dave Algren. Carrol, nongame wildlife supervisor for the Minnesota Department of Natural Resources in St. Paul, Minnesota, has compiled a decade of songbird knowledge for his book *Woodworking for Wildlife*. His pointers will guide you in construction. Dave, a woodworking hobbyist from Stillwater, Minnesota, lent a hand with the plans shown here and on the following pages. A Northwest Airlines pilot, Dave has handcrafted more than 26,000 birdhouses!

If you follow the dozen guidelines listed here, you'll guarantee yourself some of nature's finest entertainment.

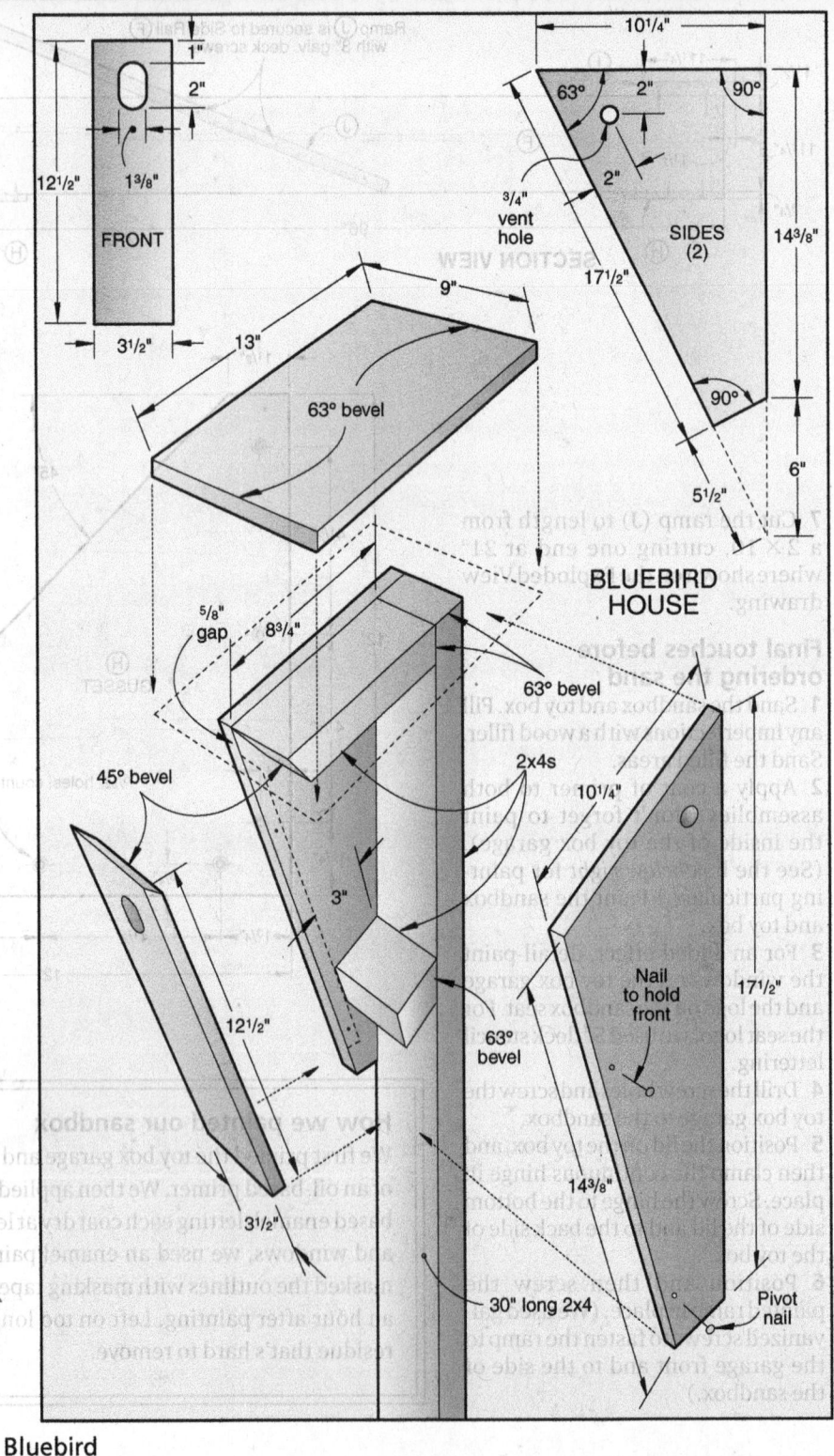

Happy building!

Bluebird

BLUEBIRD HOUSE

196   Project 3

Article adapted from WOOD magazine, issue 77, February 1995

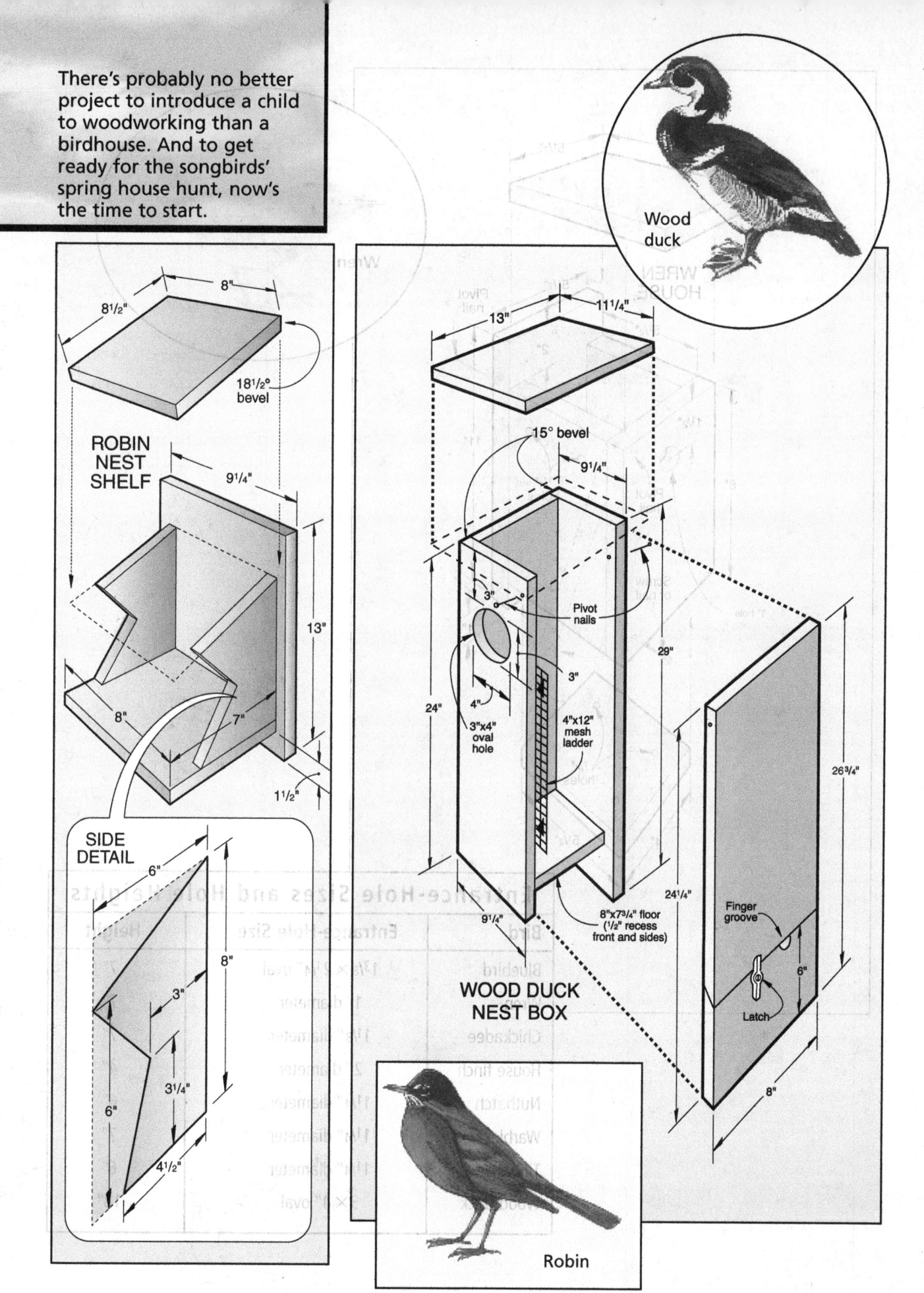

There's probably no better project to introduce a child to woodworking than a birdhouse. And to get ready for the songbirds' spring house hunt, now's the time to start.

Wood duck

ROBIN NEST SHELF

8"
8½"
18½° bevel
9¼"
13"
8"
7"
1½"

SIDE DETAIL

6"
8"
3"
3¼"
6"
4½"

WOOD DUCK NEST BOX

13"
11¼"
15° bevel
9¼"
3"
Pivot nails
3"
29"
24"
4"
3"×4" oval hole
4"×12" mesh ladder
9¼"
8"×7¾" floor (½" recess front and sides)
24¼"
Finger groove
26¾"
6"
Latch
8"

Robin

Article adapted from WOOD magazine, issue 77, February 1995

Project 3   197

WREN HOUSE

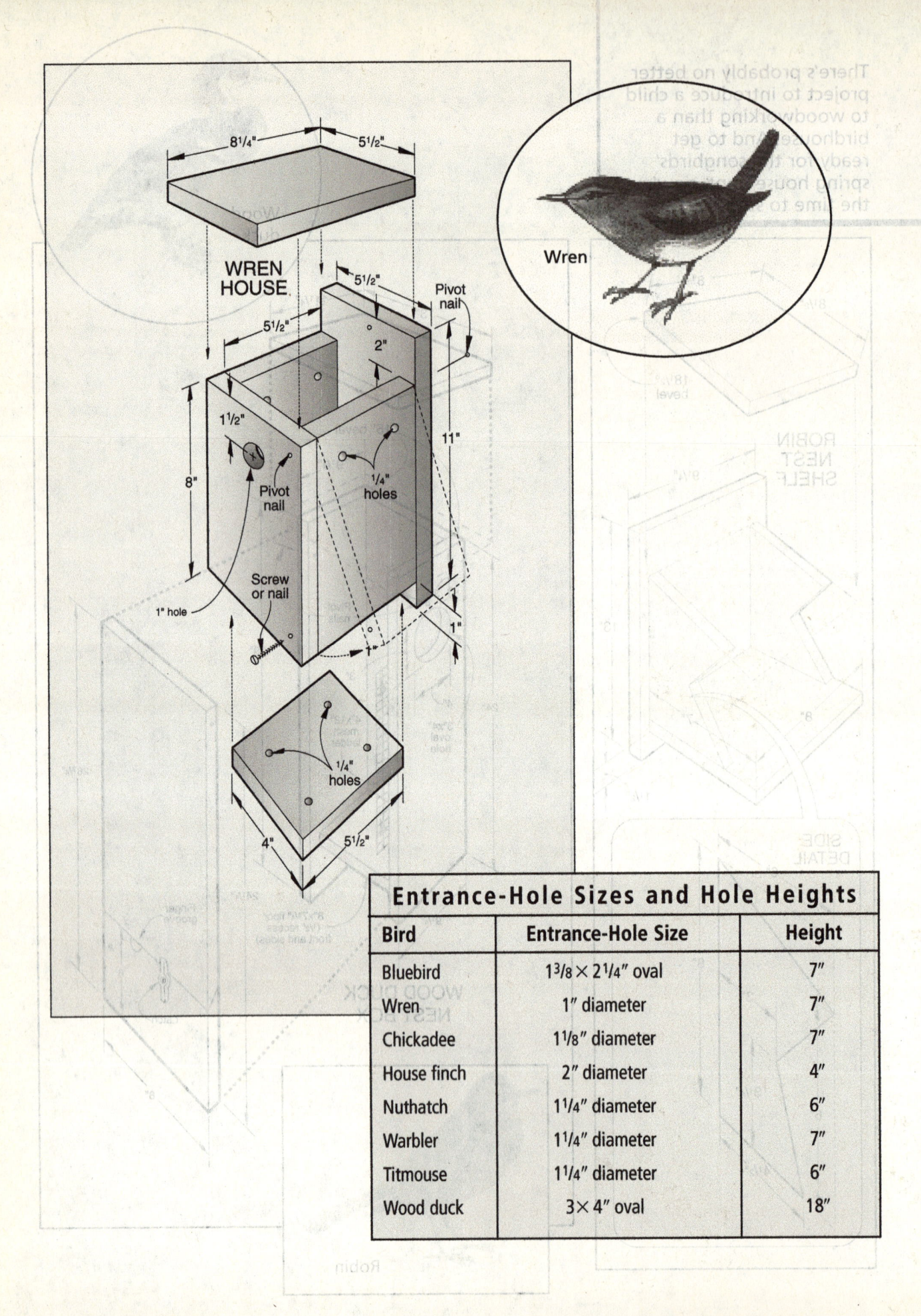

Wren

Entrance-Hole Sizes and Hole Heights		
Bird	**Entrance-Hole Size**	**Height**
Bluebird	1 3/8 × 2 1/4" oval	7"
Wren	1" diameter	7"
Chickadee	1 1/8" diameter	7"
House finch	2" diameter	4"
Nuthatch	1 1/4" diameter	6"
Warbler	1 1/4" diameter	7"
Titmouse	1 1/4" diameter	6"
Wood duck	3 × 4" oval	18"

Article adapted from WOOD magazine, issue 77, February 1995

12 IMPORTANT DO'S AND DON'TS

1 **Don't build a house just for "birds."** Build houses, nesting boxes, and other structures with specific types of birds in mind because each species has different size and entrance-hole requirements. See the chart on page 198 for dimensions and allowable entrance-hole sizes for songbird species. (A hole cut to the correct size keeps unwanted birds out. For instance, sparrows will enter holes 1¼" and larger.) Drill all holes—entrance and ventilation—at a slight upward angle to prevent rain from blowing in.

2 **Wood is the preferred material for birdhouses.** Metal does not provide heat insulation. But use only pine, cedar, redwood, or cypress—not treated wood or plywood—for functional birdhouses.

3 **Assemble cedar and redwood birdhouses with galvanized screws or resin-coated or ring-shank nails.** If you don't, the joints will eventually loosen. For pine houses, use standard fasteners.

4 **Always build the birdhouse so that the sides enclose the floor.** This keeps rain from seeping into the sidewall/floor joint. To slow deterioration of the floor, recess it ¼".

5 **The front edge of the birdhouse roof should overhang at least 2".** The overhang protects the entrance hole from rain and keeps predators from reaching in from above.

6 **So that you can clean out birdhouses semi-annually (before and after nesting season), always build them with a hinged side or roof.** Use rustproof hinges and either screw closure or a pair of roofing nails and wire to discourage raiding by raccoons.

7 **Drill at least four ⅜"-diameter drain holes in the bottom of a house (except on some special designs for bluebirds and wood duck nest boxes).** Drain holes allow rain and condensation to escape. Clear them every time you clean the house.

8 **For ventilation in all birdhouses (except duck boxes), drill at least two ⅜" holes near the top on both sides.** Wood provides great insulation, but interiors can overheat.

9 **Never put a perch on a birdhouse.** Perches encourage sparrows and European starlings, which compete with—and often kill—songbirds.

10 **Do not paint, stain, or treat with preservative the inside of a birdhouse.** You may coat the outside back of a birdhouse (the most prone to rot) with preservative, or paint the entire exterior.

11 **Firmly attach all houses to a support post, building, or tree.** If you think that cats and/or raccoons will be a problem with a post mount, discourage them with sheet-metal shields tacked to the post. Or, smear the post with grease. Wren houses can swing suspended from an eave or tree limb with a two-point suspension system.

How high to mount a birdhouse? Most songbirds nest in a range from 4–15' above the ground. Remember, though, that you need to reach it for cleaning. And remember to provide shade for at least part of the day.

12 **To avoid territorial fights, space houses for songbirds at least 20' apart.** Space bluebird houses 100 yards apart. Purple martins and wildfowl, such as wood ducks, don't defend their territories.

Article adapted from WOOD magazine, issue 77, February 1995

COUNTRY

This simply styled furniture piece proves that you don't have to spend a fortune to spruce up your favorite outdoor area.

Note: For our chair, we hand-picked fir 2× stock. Pine, spruce, or redwood also will work well. If you have trouble locating straight and uncupped stock, edge-join narrower pieces to width. For joints that will stand up to the extremes of Mother Nature, use water-resistant glue, slow-set epoxy, or resorcinol glue.

Build the chair ends first

1 Cut the ends (A) to 20¾" long from 2 × 12 or edge-joined stock.

2 Transfer the full-sized heart half-pattern onto a piece of heavy paper or poster board. Cut the full-sized template to shape.

3 Position the template, and trace the heart outline on all four end pieces (A) where located on the End View drawing. Cut the marked outlines to shape on the bandsaw or with a jigsaw, and then drum-sand the cut edges to remove the saw marks.

4 Clamp each matching pair of 2×12s together, heart edge to edge, with the top and bottom edges flush. Now, using the dimensions on the Exploded View drawing, mark the two dowel-hole locations on one face. Remove the clamps. Using a square, transfer the lines to the inside edge of each end piece.

5 Check that you're square to the edge, and bore ¼" holes 1½" deep centered from edge to edge where marked. (We used an electric drill and a spade bit.)

6 Set a stop, and crosscut four pieces of ¾" oak dowel stock to 3⅜" long. For ease of insertion, sand a chamfer on each end of each dowel. (We formed our chamfers on a belt sander.)

7 Cut four ⅜"-thick spacers. Glue, dowel, and clamp both chair ends together, placing the ⅜" spacers between the end pieces (A) for a consistent ⅜" gap. Save the spacers; you'll use them when joining the seat and back pieces later.

8 Using trammel points, swing an arc to mark the 15⅜" radius on the bottom end of each chair assembly where dimensioned on the End View drawing. Cut the arcs to shape.

9 Sand a slight round-over on all edges of each chair end assembly.

Now, for the seat

1 Cut the two seat pieces (B, C) to length. (We ripped the two pieces to width from 2 × 12 stock.)

2 Rout a ½" round-over along the top front edge of the seat front piece (B).

200   Project 4

Article adapted from WOOD magazine, issue 69, April 1994

CHAIR

Affordable, easy-to-build seating for your deck or patio

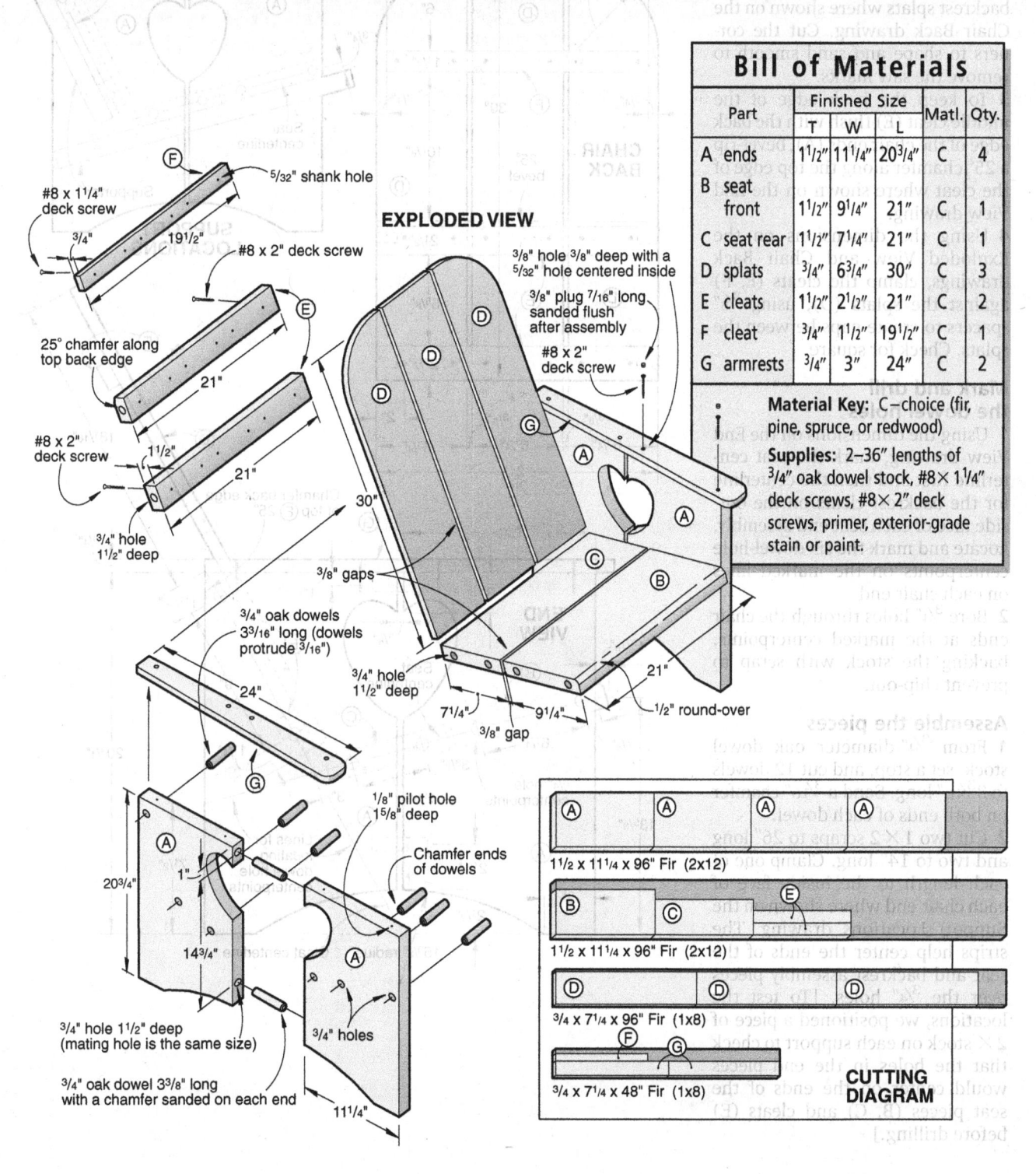

EXPLODED VIEW

Bill of Materials

Part	T	W	L	Matl.	Qty.
A ends	1 1/2"	11 1/4"	20 3/4"	C	4
B seat front	1 1/2"	9 1/4"	21"	C	1
C seat rear	1 1/2"	7 1/4"	21"	C	1
D splats	3/4"	6 3/4"	30"	C	3
E cleats	1 1/2"	2 1/2"	21"	C	2
F cleat	3/4"	1 1/2"	19 1/2"	C	1
G armrests	3/4"	3"	24"	C	2

Material Key: C—choice (fir, pine, spruce, or redwood)

Supplies: 2–36" lengths of 3/4" oak dowel stock, #8 × 1 1/4" deck screws, #8 × 2" deck screws, primer, exterior-grade stain or paint.

CUTTING DIAGRAM

1 1/2 x 11 1/4 x 96" Fir (2x12)
1 1/2 x 11 1/4 x 96" Fir (2x12)
3/4 x 7 1/4 x 96" Fir (1x8)
3/4 x 7 1/4 x 48" Fir (1x8)

Article adapted from WOOD magazine, issue 69, April 1994

Project 4   201

CHAIR

Build the backrest next

1 Cut the backrest splats (D) and backrest cleats (E, F) to size.

2 Mark a 6" radius on two of the backrest splats where shown on the Chair Back drawing. Cut the corners to shape and sand smooth to remove the saw marks.

3 To keep the back edge of the middle cleat (E) flush with the back edge of the chair ends (A), bevel-rip a 25° chamfer along the top edge of the cleat where shown on the End View drawing.

4 Using the dimensions on the Exploded View and Chair Back drawings, clamp the cleats (E, F) against the splats (D), using 3/8" spacers to create gaps between the splats. Check for square.

Mark and drill the dowel holes

1 Using the dimensions on the End View drawing, mark the seat centerline first and then the centerline for the backrest cleats on the outside face of each seat end assembly. Locate and mark the six dowel-hole centerpoints on the marked lines on each chair end.

2 Bore 3/4" holes through the chair ends at the marked centerpoints, backing the stock with scrap to prevent chip-out.

Assemble the pieces

1 From 3/4"-diameter oak dowel stock, set a stop, and cut 12 dowels to 3 3/16" long. Sand a 3/16" chamfer on both ends of each dowel.

2 Cut two 1 × 2 scraps to 26" long and two to 14" long. Clamp one of each length to the inside face of each chair end where shown on the Support Locations drawing. The strips help center the ends of the seat and backrest assembly pieces over the 3/4" holes. [To test the locations, we positioned a piece of 2× stock on each support to check that the holes in the end pieces would center on the ends of the seat pieces (B, C) and cleats (E) before drilling.]

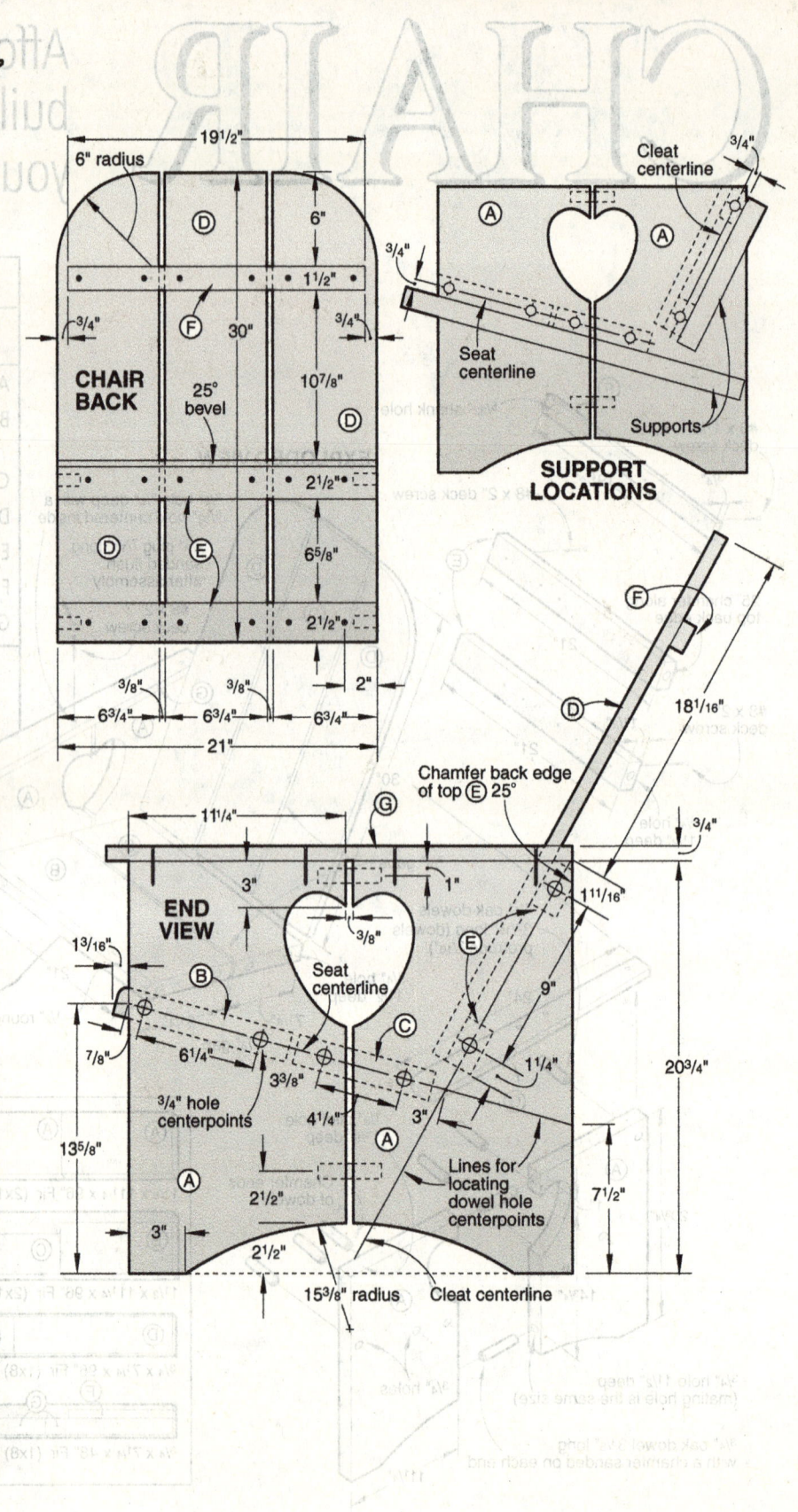

202   Project 4

Article adapted from WOOD magazine, issue 69, April 1994

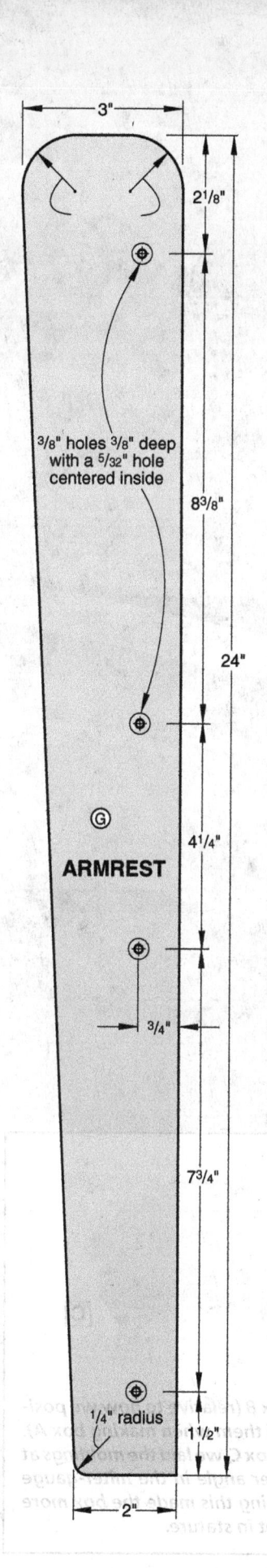

3 With a helper, position the seat pieces where located on the End View drawing. Slip the ³⁄₈" spacers between the pieces for a consistent gap. Clamp the seat pieces firmly between the chair ends.

4 Chuck a ¾" spade bit into a portable electric drill. Using the previously bored holes in the end sections as guides, bore a pair of 1½"-deep holes squarely into each seat piece end. As soon as you've bored the first hole, insert one of the 3³⁄₁₆"-long dowels into the hole to help steady the seat piece for boring the next hole. Do not insert the dowel more than ½" into the seat piece; you may have trouble removing it if you insert it all the way.

5 Repeat the procedure to position and drill the ¾" holes in both ends of the backrest cleats (E).

6 Remove one of the 3³⁄₁₆"-long dowels. With a small brush, coat the inside of the hole with glue. To prevent marring the chamfered dowel end, use a rubber-tipped mallet to slowly drive the dowel into the hole. Drive the dowel until just the chamfered end protrudes. Be careful not to drive the dowels too far—they're almost impossible to back out. Immediately wipe off any excess glue. Repeat for each dowel. Let the glue dry and then remove the clamps.

Now, let's add the armrests

1 From ¾" stock, cut two pieces to 3 × 24" for the armrests (G).

2 Using the dimensions on the armrest drawing, mark the profile on one piece and cut it to shape. Use the first piece as a template to mark the shape onto the second armrest, and then cut it to shape.

3 Mark the hole centerpoints. Drill and counterbore the holes.

4 Screw the armrests to the tops of the end assemblies (A).

5 Plane or resaw a piece of stock to ⁷⁄₁₆" thick, and use a plug cutter to cut ³⁄₈"-diameter plugs. Plug the holes, and sand the plugs flush with the top of each armrest.

Sand, paint, and sit a spell

1 Sand the entire chair, sanding a slight round-over on all edges.

2 If you decide to paint your chair, an oil-based enamel or water-based latex will provide adequate protection. Regardless of your paint selection, be sure to apply a prime coat that's compatible with the top coat. Also, be sure to apply several coats to the porous end grain.

For a more natural look, finish the chair with an exterior house stain, and then apply several coats of spar varnish.

Article adapted from WOOD magazine, issue 69, April 1994

A tale of three boxes

Believe it or not, these boxes were made from crown molding of the same size and profile, but with slight building modifications. Box A is made of red oak with an oak handle. Boxes B and C, made of cherry, have cardinal wood and Corian handles, respectively.

We made boxes A and B in the same way, except we positioned the crown moldings upside down in the jigs when cutting the parts for box B (relative to how we positioned them when making box A). With box C, we laid the moldings at a flatter angle in the miter-gauge jig. Doing this made the box more upright in stature.

204   Project 5

Article adapted from WOOD magazine, issue 104, March 1998

Crown Molding Boxes

They're shapely, stately, and oh so easy to make

They look like the work of an artist, with their elegant curves and classic proportions. Actually, you can make these delightful boxes with crown molding available at any home center. The trick is in the jigs, and the jigs aren't complicated at all. So, just spend a little time jigging up, and you'll soon be turning out a variety of impressive boxes. Don't worry—no mathematics or protractors required.

This technique began to hatch in the mind of assistant design editor Jan Svec several years ago. A fellow employee at a millwork shop would take scraps of molding, miter them, and make boxes with vertical sides.

Not bad, thought Jan. Not bad for a beginning.

The next step came when he toured historic homes in Philadelphia, searching for project design ideas for WOOD® magazine. "In one house, I saw a tea box with a molding profile inside and out," Jan said. That would be a very complicated project, but what about a box that's fancy on the outside and plain on the inside? He realized that crown molding would do the trick.

However, part of the beauty of this tea box came from its angled sides, so Jan experimented with ways to cut compound miters without complicated math or fancy equipment. In fairly short order, he came up with the two jigs shown on page 206.

Why two jigs instead of one? It has to do with the limited height of a table saw blade and the need for a high-profile box topped by a low-profile lid. Make them, try them, and you'll see. You'll soon cut these compound miters without ever having to figure an angle in degrees.

You can use the jigs as shown to make boxes of any width and length. They will handle molding up to 4⅝" wide. As you can see in the sidebar *left*, changing the position of the workpieces in the jigs dramatically changes the appearance of the box. In this article, we'll show you how to build a box like the one in the large photo. Then you can experiment with other styles.

Continued

Article adapted from WOOD magazine, issue 104, March 1998

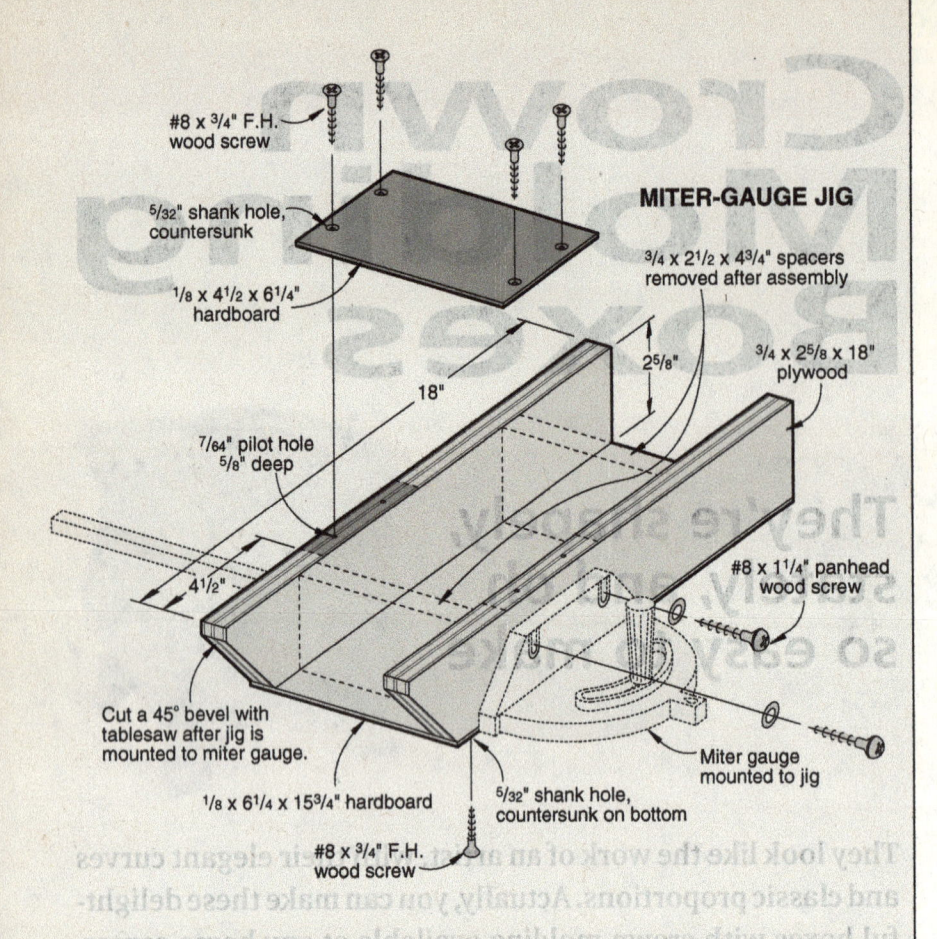

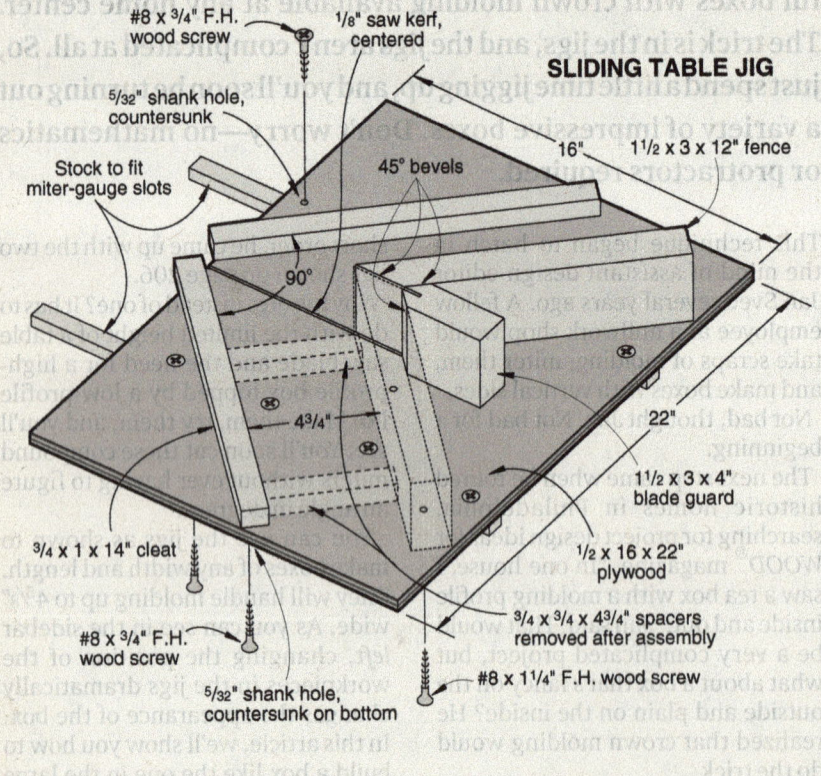

Crown Molding Boxes

Ready to try one? First, cut the four sides

Our finished box will measure about 8" wide by 10" long. From a piece of 4¼"-wide crown molding, cut two pieces 8½" long and two pieces 10½" long. Cut them in the sequence that they'll follow around the box—side, end, side, end—and number each one near the bottom edge. (This way, the grain will seem to "run" continuously around the box.)

Set your table saw blade to 45°, or just a hair over, to make sure that the outside points of the miters will be tight. Slip the miter-gauge jig into the right-hand slot of the table saw, and load one of the four pieces into it. Lean the molding against the rear fence at the steepest angle that will still allow the blade to cut through it, then measure the distance between the lower edge and the front fence. Cut a spacer to that width, and put it between the workpiece and the fence, as shown.

The edge that sits higher in the jig will become the top edge of the finished box; the lower edge will be at the bottom. Double-check that before making each cut. Now, with the workpiece wedged firmly between spacer and fence, cut a miter. Do the same at one end of each of the four pieces.

To miter the opposite ends, flip one of the shorter pieces so it's leaning against the front fence, and place the spacer between the workpiece and the rear fence. Line it up with the blade, and clamp a stopblock at the other end. You don't have to measure a thing. Cut that miter, take the piece out, and cut the second short piece exactly the same way, without moving the stopblock.

Remove the stopblock and follow the same procedure for the longer pieces. Tape the four pieces together to check the fit.

206 Project 5

Article adapted from WOOD magazine, issue 104, March 1998

Adding feet and a bottom won't take you long

Along the bottom edge of each workpiece, measure in 1" from each end and make a mark. Draw a line parallel to the bottom edge and 1" up from that edge. With a compass set to a 1" radius, scribe a curve up to the line.

Use a scroll saw or bandsaw to cut close to that line. Install a 2" drum in your spindle sander or on your drill press, and sand to the line as shown in the photo below.

As shown in the drawing (bottom), use a sliding bevel gauge and straightedge to determine the angle for the kerf that holds the box bottom. Again, the number of degrees isn't important. Just set the sliding bevel gauge and use it to set your table saw blade.

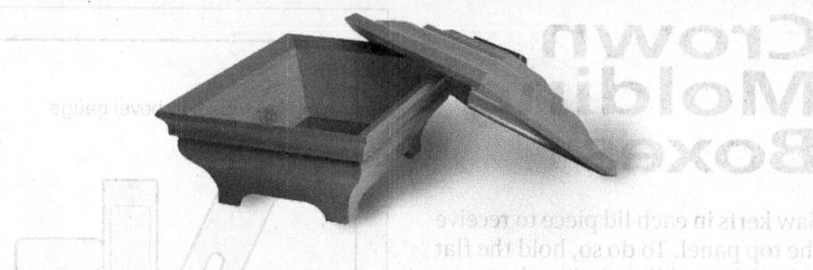

With the feet of each piece against the rip fence and the molding profile facing up, saw a 1/8"-wide kerf about 1/8" above the leg cutout and 3/16" deep at its shallow side. Do this on all four pieces. Measure the lengths of those kerfs to find the dimensions of the bottom, and cut a piece of 1/8" plywood or solid wood to fit.

After test-fitting, glue together the sides, ends, and bottom of the box. Hold them together with masking tape and a band clamp until the glue dries. Make sure the top edges are lined up at the corners. It's easier to sand away imperfections if they're on the bottom.

Top off your beautiful box with a matching lid

As you choose molding stock for the lid, keep in mind that the lid pieces cannot meet in the center of the box. That's because there must be a panel at least 1" wide in the center of the lid for mounting a handle. For the box in our example, 3 1/2"-wide crown molding will do the trick. If the box was wider, the 4 1/4"-wide stock used for the base might work for the lid as well.

From the 3 1/2" crown molding, cut two pieces 8 1/2" long and two pieces 10 1/2" long. As before, cut in a side-end-side-end sequence and number the pieces.

Also cut a scrap of that molding 1" long and tape it to the box so that its bottom edge fits snugly against the bevel of the box. Lay a straight piece of scrapwood across the box and mark the angle of your scrap molding on it. (See the drawing at bottom.) Set the sliding bevel gauge to that angle, and use it to set the table saw blade, with the handle of the gauge held against the rip fence as shown on page 208.

Continued

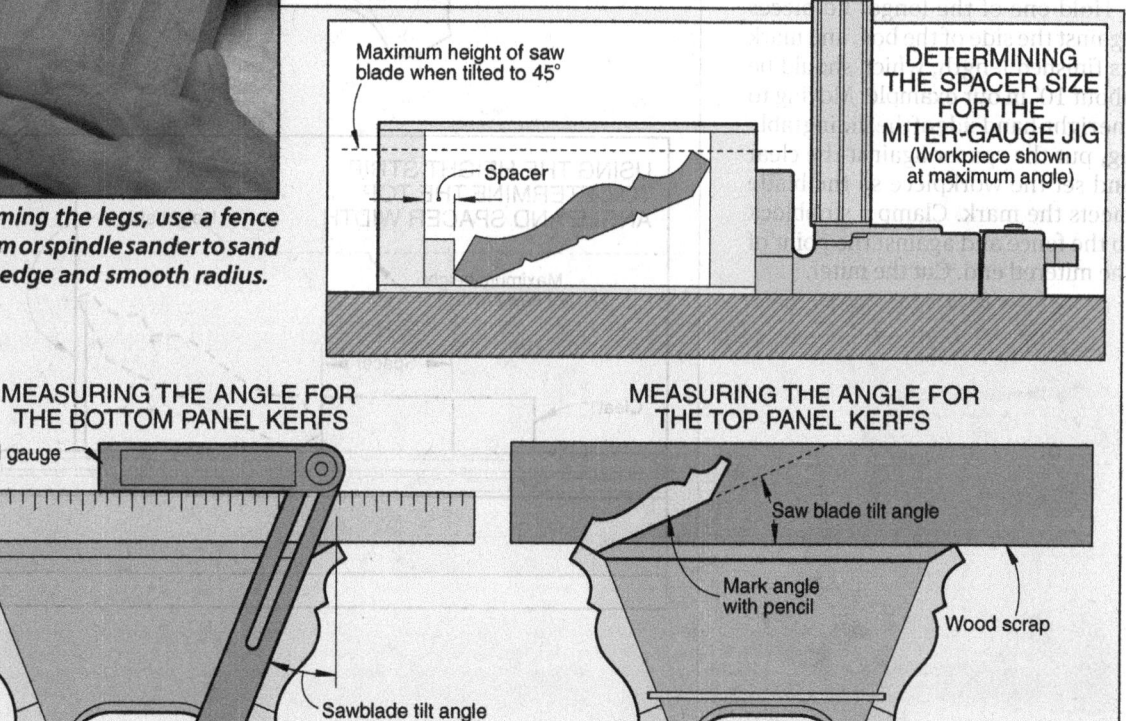

When forming the legs, use a fence with a drum or spindle sander to sand a straight edge and smooth radius.

DETERMINING THE SPACER SIZE FOR THE MITER-GAUGE JIG
(Workpiece shown at maximum angle)

MEASURING THE ANGLE FOR THE BOTTOM PANEL KERFS

MEASURING THE ANGLE FOR THE TOP PANEL KERFS

Article adapted from WOOD magazine, issue 104, March 1998

Crown Molding Boxes

Saw kerfs in each lid piece to receive the top panel. To do so, hold the flat side of the molding against the fence, and cut a slot 1/8" wide and 3/16" deep, at least 1/8" from the edge that will be the highest part of the lid. See the photo at right.

Place the sliding table jig in the slots of your table saw, and set the blade at exactly 90°. To find the correct angle for the miter cut, again use a piece of scrap molding 1" long taped to the box. With a straightedge across the box, measure as shown in the illustration (right middle). Cut a piece of scrap to a length that equals A-B+C. Hold this height strip flat against the left fence of the jig, and draw a line along the top edge.

Tilt one of the molding pieces against the fence so that it just covers the line, and measure the gap between the workpiece and the cleat. Cut a spacer to fit that gap. See the drawing at right bottom. Cut a miter at one end of each of the four pieces, each time pressing the piece firmly between the fence and spacer.

Hold one of the longer lid pieces against the side of the box, and mark its finished length, which should be about 10" in our example. Moving to the right-hand side of the sliding table jig, put the spacer against the cleat and set the workpiece so the blade meets the mark. Clamp a stopblock to the fence and against the point of the mitered end. Cut the miter.

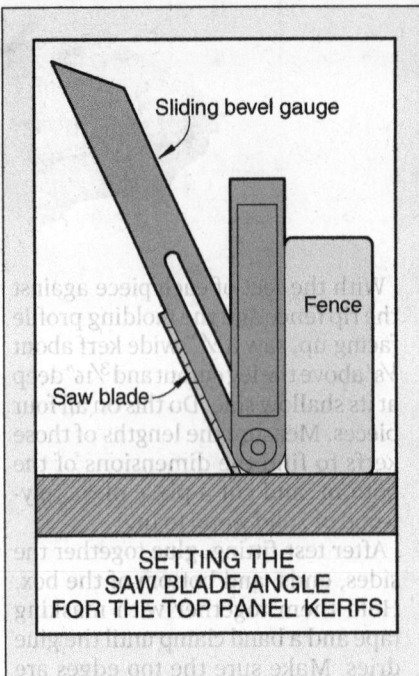

SETTING THE SAW BLADE ANGLE FOR THE TOP PANEL KERFS

Hold the back side of the molding against your table saw's fence to cut the kerf that holds the panel in the center of the lid.

MEASURING FOR THE HEIGHT STRIP TO SET THE TOP SLOPE

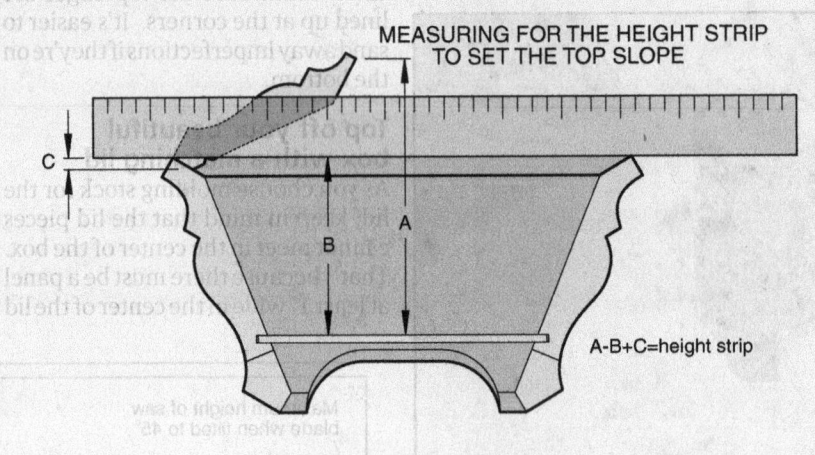

A-B+C=height strip

USING THE HEIGHT STRIP TO DETERMINE THE TOP ANGLE AND SPACER WIDTH

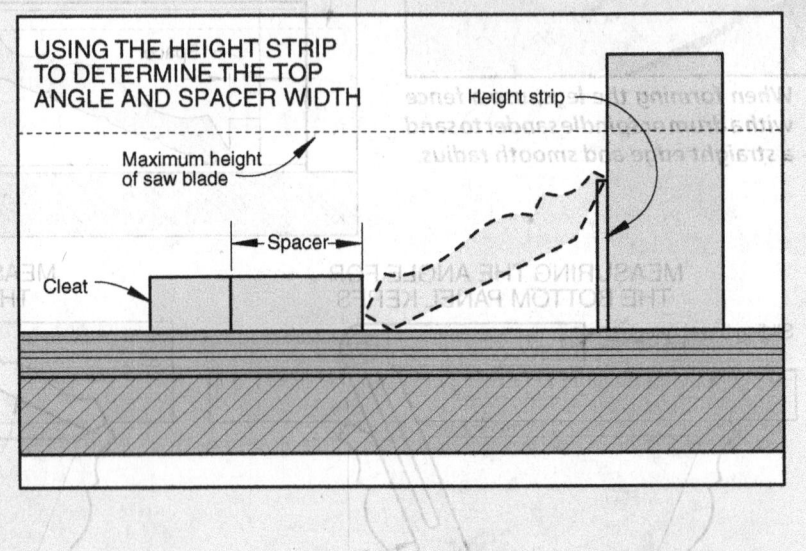

208 Project 5

Article adapted from WOOD magazine, issue 104, March 1998

A few facts about crown molding

You can buy crown molding in many sizes and species. Shown here are some of the moldings we worked with during the making of this article. Each one will produce a box with a slightly different look.

Commonly available species include oak, cherry, and poplar, and range in width from 3½" to 7¼". (If you use moldings wider than 4⅝", you will need to upsize our jigs to accommodate their greater width. Your larger jigs will work just like the ones shown in this article.) We used economically priced poplar for the painted boxes.

When you go shopping, check a couple of sources for selection and price, and inspect the molding's surface carefully. Some mills turn out glass-smooth cuts, and some leave noticeable chatter marks that can be tough to sand out of an elaborate profile.

Also, some mills cut grooves in the back side of the molding and some mills don't. If you prefer a smooth interior, be sure to flip the molding over and check the back before buying.

A wide range of available crown moldings means you can make an infinite variety of boxes.

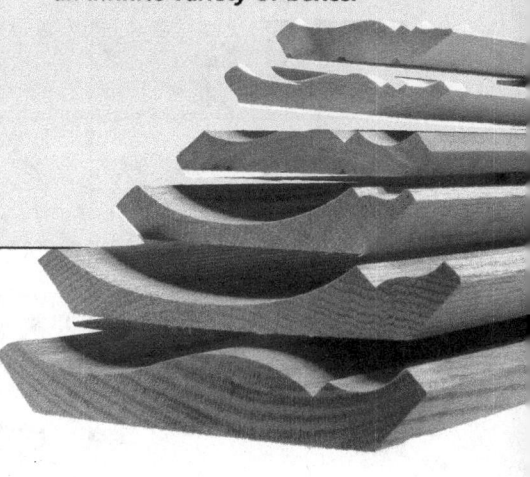

Attach four cleats at right angles to each other to hold the lid pieces in position on a plywood clamping platform.

Apply glue to the mitered edges and squeeze them together with the help of some scrap stock and clamps.

Repeat the procedure for the other long piece, and follow the same steps for the two shorter pieces. Again, the bottom edge of the lid goes at the bottom when cutting. Cut a piece of ⅛" plywood or solid wood to fit the lid kerfs.

To make a clamping jig for the lid, screw two pieces of scrap to a piece of plywood at right angles, hold the lid together, and set it into that corner. Screw two more scrap pieces into the plywood so that they're tight against the lid as shown in the photo (left middle).

Take the lid out, glue the miters, and replace the lid assembly into the clamping jig. A board or two on top, held down by clamps, will force the miters snugly together as shown (left bottom).

For a handle, we suggest a "fin" made of ¼–½" stock, cut to the length of the lid panel and about ½–1" wide. This would be a fine time to use a piece of exotic scrapwood or Corian. Attach the handle with brass screws from the bottom side of the panel.

Apply the finish of your choice. Paint works well for woods lacking showy grain patterns, or if you want to show off the grain, apply several coats of oil and top it off with paste wax.

Article adapted from WOOD magazine, issue 104, March 1998

WORKBENCH

with laminated maple top

with plywood top

Note: We show two versions of the top for this workbench: an easy plywood top (above) and the traditional laminated style (left). The following step-by-step directions explain how to construct the laminated top. To build the plywood top, simply glue and screw three pieces of ¾"-thick plywood face to face (we used birch), and then trim to 18¾×44½". Band the top with ¾"-thick mitered strips of solid 2¼"-wide birch stock.

Let's begin with the top

1 Rip and crosscut 26 splines ¹¹⁄₁₆×48" from ¼"-thick plywood. Next, rip and crosscut 27 lengths of ¾"-thick stock (we used maple) to 2⁵⁄₁₆×48" to make initially oversized pieces for the top (A). Mount a ¼" dado blade in your table saw, and set the rip fence ¾" from the inside edge of the blade. Set a ⅜" depth of cut, and make test cuts in scrap wood to ensure that the width

A sturdy, dependable workbench belongs at the center of every hardworking home workshop. This heavy-duty, versatile model features a pair of vises that provide a variety of ways to hold workpieces. You'll also like the knock-down construction, allowing you to easily move the bench. In addition, you can choose from two tops—an indestructible laminated one, or a simpler plywood version. Both will give you years of solid service.

210 **Project 6** Article adapted from WOOD magazine, Weekend Woodworking Projects, September 1995

1 EXPLODED VIEW

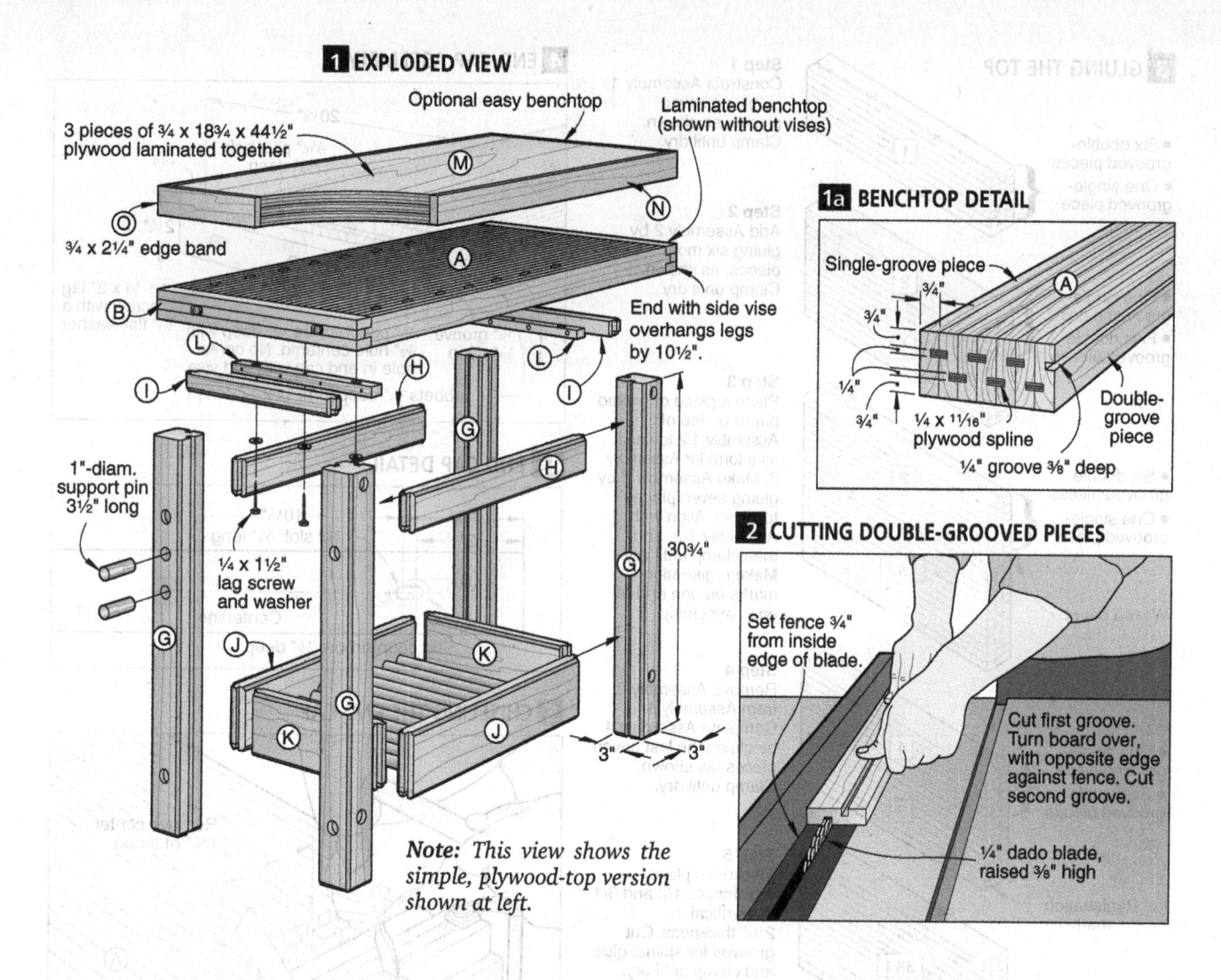

Labels in diagram:
- 3 pieces of ¾ x 18¾ x 44½" plywood laminated together
- Optional easy benchtop
- Laminated benchtop (shown without vises)
- ¾ x 2¼" edge band
- End with side vise overhangs legs by 10½"
- 1"-diam. support pin 3½" long
- ¼ x 1½" lag screw and washer
- 30¾"
- 3" 3"
- Note: This view shows the simple, plywood-top version shown at left.

1a BENCHTOP DETAIL

- Single-groove piece
- ¾"
- ¾"
- ¼"
- ¾"
- ¼ x 11/16" plywood spline
- Double-groove piece
- ¼" groove ⅜" deep

2 CUTTING DOUBLE-GROOVED PIECES

- Set fence ¾" from inside edge of blade.
- Cut first groove. Turn board over, with opposite edge against fence. Cut second groove.
- ¼" dado blade, raised ⅜" high

---

of the groove matches your plywood splines. Also, make certain that the spline does not bottom out in the grooves, preventing the top pieces from fitting together tightly.

2 Referring to Drawing 1a accompanying Drawing 1, cut single grooves in four pieces. Then, cut double grooves in 23 pieces, as shown in Drawing 2.

3 Glue the top pieces together, following the sequence described on Drawing 3.

Note: In Step 3, you use the completed Assembly 1 & 2 as a form to begin the glue-up of Assembly 3. This way, any irregularity in the edge of Assembly 1 & 2 will transfer to Assembly 3, ensuring a good fit when you join the two assemblies. Double-check the flatness of the top when you glue the two assemblies together. When the glue is dry, square the ends of the top, trimming the assembly to a finished length of 44".

Now, make the end caps, and-mount the vises

1 Rip and crosscut six 2×21" pieces from ¾"-thick stock to make initially oversized blanks for the end caps (B). Glue and clamp two sets of three pieces face to face. When the glue is dry, joint one edge of each lamination, then rip and crosscut to finished size as dimensioned in Drawing 4.

2 Mount a ¾" dado blade in your table saw, and cut a ¾×¾" groove centered in one edge of the end cap lamination where shown. Add an auxiliary wooden fence to your rip fence, adjust the fence, and cut ¾×¾" rabbets along both ends of the top.

Note: These rabbets will produce a ¾×¾" tenon centered in both ends of the top.

Raise the blade in small increments to get a perfect fit of the tenon into the end cap. Cut a ¾"-wide stopped groove 11" long into the bottom of the workbench, where dimensioned on Drawing 6.

3 Remove stock from the end cap to make room for the end vise, as shown in Drawing 5. Next, trim the tenon as shown.

4 Drill the counterbores, slots, and center hole into the end caps where dimensioned on Drawing 4a. The end cap that will be used next to the end vise does not have a center hole. Hold each end cap in position, and use the hole and slots as guides to drill 3/16" pilot holes 1½" deep into the ends of the top. Be sure that you center pilot holes in the slots. Attach caps with lag screws and washers.

Note: The slots in end caps (B) allow the top to expand and contract with seasonal changes in the wood's moisture content.

Article adapted from WOOD magazine, Weekend Woodworking Projects, September 1995

Project 6   211

3 GLUING THE TOP

- Six double-grooved pieces
- One single-grooved piece
} 1

- One single-grooved piece
- Five double-grooved pieces
} 2
1

- Six double-grooved pieces
- One single-grooved piece
} 3
1&2

Waxed paper

- One single-grooved piece
- Six double-grooved pieces
} 4
3

Registration mark

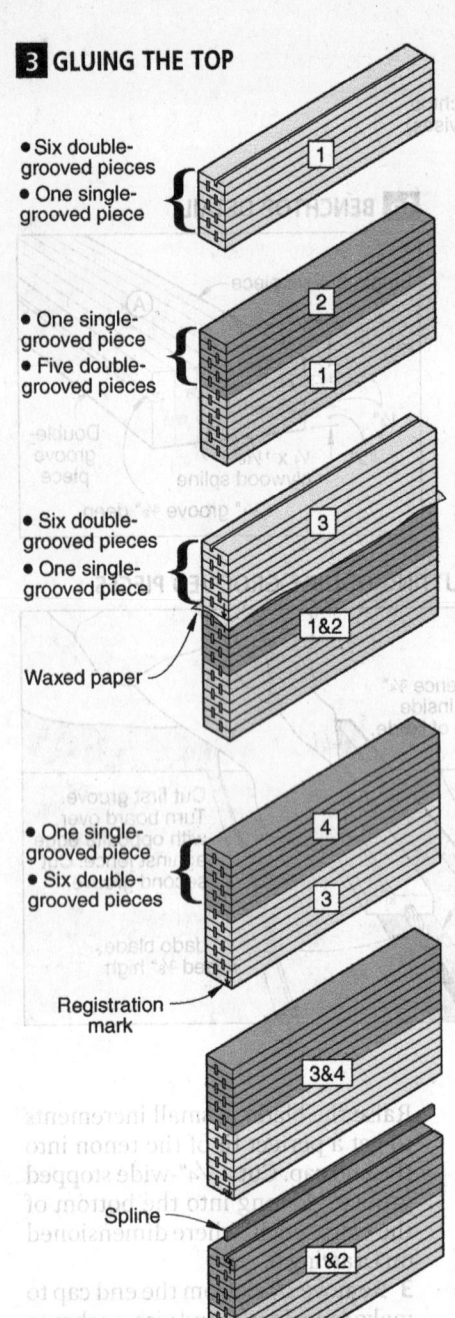

Spline

3&4
1&2

Step 1
Construct Assembly 1 by gluing seven pieces, as shown. Clamp until dry.

Step 2
Add Assembly 2 by gluing six more pieces, as shown. Clamp until dry.

Step 3
Place a piece of waxed paper on top of Assembly 1/2 to use it as a form for Assembly 3. Make Assembly 3 by gluing seven pieces together. Align with Assembly 1/2, and then clamp until dry. Make registration marks on one end of each assembly.

Step 4
Remove Assembly 3 from Assembly 1/2. Complete Assembly 4 by gluing the last seven pieces, as shown. Clamp until dry.

Step 5
Thickness-plane assemblies 1/2 and 3/4 to identical 2¼" thickness. Cut grooves for spline; glue and clamp until dry.

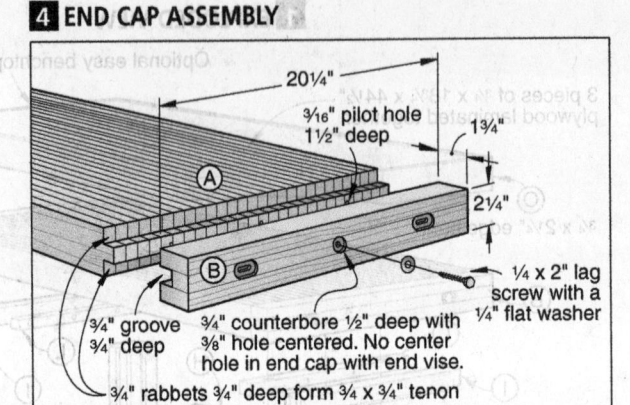

4 END CAP ASSEMBLY

20¼"
³⁄₁₆" pilot hole 1½" deep
1¾"
2¼"
¼ x 2" lag screw with a ¼" flat washer
¾" groove ¾" deep
¾" counterbore ½" deep with ⅜" hole centered. No center hole in end cap with end vise.
¾" rabbets ¾" deep form ¾ x ¾" tenon

4a END CAP DETAIL

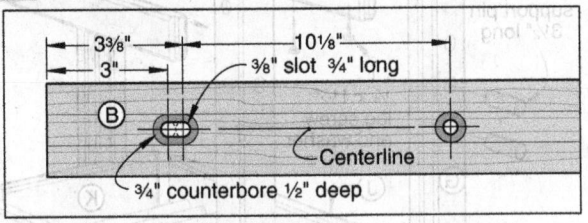

3⅜"
3"
10⅛"
⅜" slot ¾" long
Centerline
¾" counterbore ½" deep

5 CHISELING THE END CAP

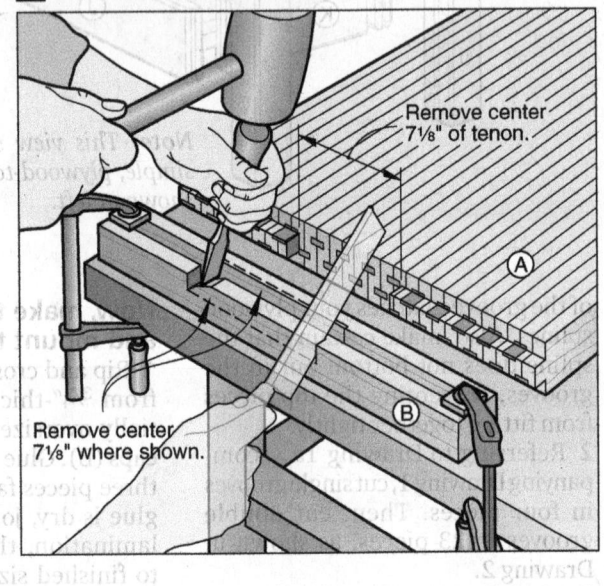

Remove center 7⅛" of tenon.
Remove center 7⅛" where shown.

5 Mark the centerpoints of the bench-dog holes, where dimensioned on Drawing 7. Next, drill the holes, using a portable-drill guide to keep the holes vertical.

6 Rip and crosscut ¾"-thick stock for the side-vise spacer (C) and the end-vise spacer (D). These parts are shown on Drawing 10. Make the radius cuts on the ends of the end-vise spacer with a jigsaw. Next, drill ½" blade-start holes, and then jigsaw the openings. Drill countersunk holes where shown. Remove the patterns, and then screw the parts into position where shown on Drawing 6.

7 Mount the side vise, where shown on Drawing 6. Using the mounting flanges in the vise's base as guides, drill ¼" pilot holes 1½" deep through the side-vise spacer (C) and into the top (A). Use a socket wrench to drive the lag screws. Rip and crosscut the side-vise cheek (E) from ¾"-thick stock. Position it within the jaw of the vise, flush with the top's end and upper surface. Close the vise to hold the end-vise cheek in position; attach with lag screws.

8 Rip and crosscut three oversized pieces to 3¼×21" from ¾"-thick stock for the end-vise cheek (F). Glue and clamp them together face to face. When dry, joint one edge, and then rip and crosscut to finished size. Double-check the finished length of the end-vise cheek against the actual width of your workbench top. Referring to

212 **Project 6**

Article adapted from WOOD magazine, Weekend Woodworking Projects, September 1995

6 TOP EXPLODED VIEW
(shown upside down)

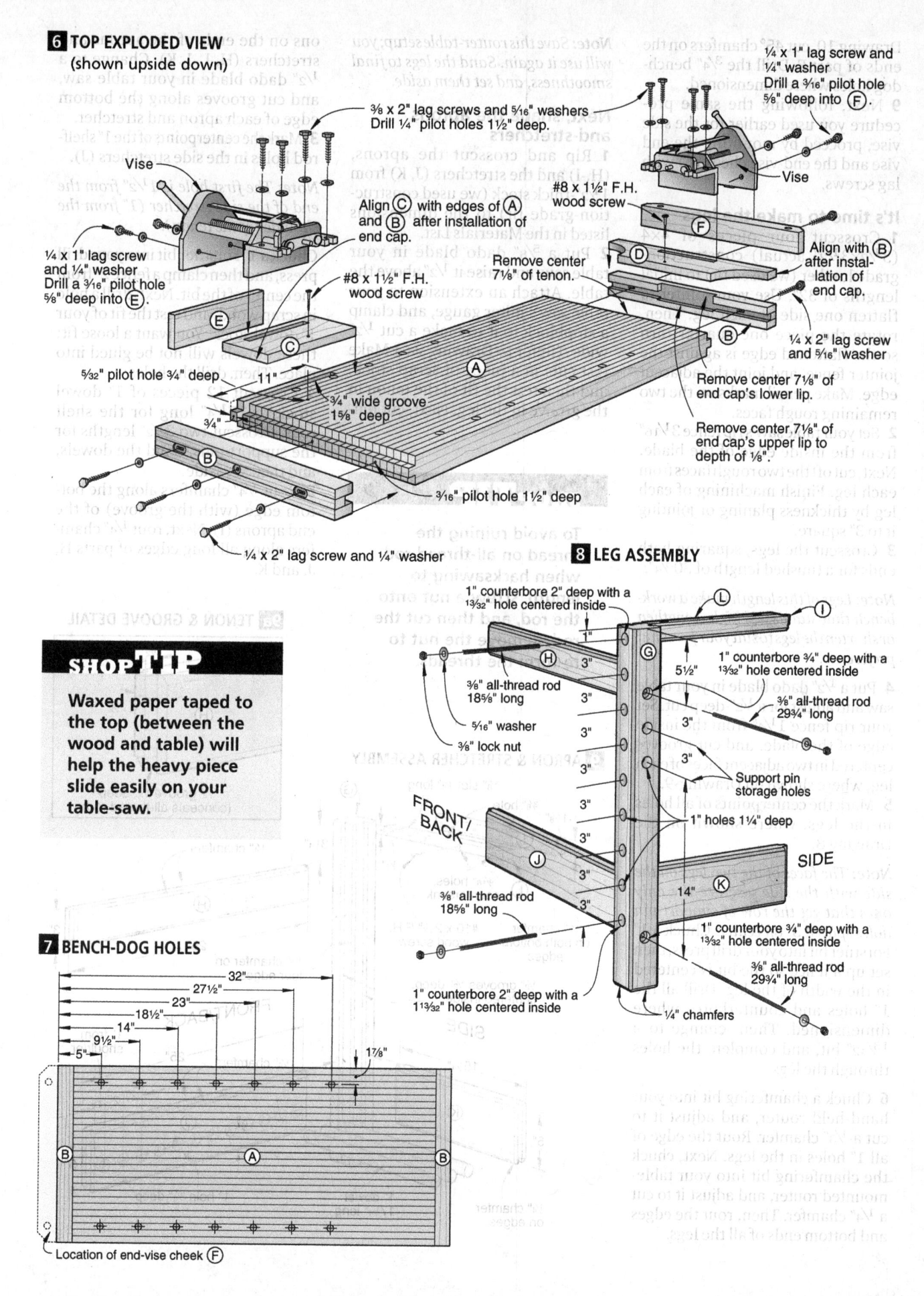

3/8 x 2" lag screws and 5/16" washers
Drill 1/4" pilot holes 1 1/2" deep.

1/4 x 1" lag screw and 1/4" washer
Drill a 3/16" pilot hole 5/8" deep into F.

Vise

Align C with edges of A and B after installation of end cap.

#8 x 1 1/2" F.H. wood screw

Remove center 7 1/8" of tenon.

Align with B after installation of end cap.

1/4 x 1" lag screw and 1/4" washer
Drill a 3/16" pilot hole 5/8" deep into E.

#8 x 1 1/2" F.H. wood screw

1/4 x 2" lag screw and 5/16" washer

Remove center 7 1/8" of end cap's lower lip.

Remove center 7 1/8" of end cap's upper lip to depth of 1/8".

5/32" pilot hole 3/4" deep

11"

3/4"

3/4" wide groove 1 5/8" deep

3/16" pilot hole 1 1/2" deep

1/4 x 2" lag screw and 1/4" washer

SHOP TIP
Waxed paper taped to the top (between the wood and table) will help the heavy piece slide easily on your table-saw.

8 LEG ASSEMBLY

1" counterbore 2" deep with a 1 13/32" hole centered inside

1" counterbore 3/4" deep with a 1 13/32" hole centered inside

3/8" all-thread rod 18 5/8" long

5/16" washer

3/8" lock nut

3/8" all-thread rod 29 3/4" long

Support pin storage holes

1" holes 1 1/4" deep

FRONT/BACK

SIDE

3/8" all-thread rod 18 5/8" long

1" counterbore 2" deep with a 1 13/32" hole centered inside

1" counterbore 3/4" deep with a 1 13/32" hole centered inside

3/8" all-thread rod 29 3/4" long

1/4" chamfers

7 BENCH-DOG HOLES

32"
27 1/2"
23"
18 1/2"
14"
9 1/2"
5"

1 7/8"

Location of end-vise cheek F

Article adapted from WOOD magazine, Weekend Woodworking Projects, September 1995

Drawing 10, cut 45° chamfers on the ends of part F. Drill the ¾" bench-dog holes where dimensioned.

9 Next, following the same procedure you used earlier for the side vise, proceed by mounting the end vise and the end-vise cheek (F) with lag screws.

It's time to make the legs

1 Crosscut four pieces of 4×4 (3½×3½" actual) construction-grade lumber (we used fir) to initial lengths of 32". Use your jointer to flatten one side of each leg. Then, rotate the piece one-quarter turn so the flattened edge is against the jointer fence, and joint the adjacent edge. Make pencil marks on the two remaining rough faces.

2 Set your table saw's rip fence 3 1/16" from the inside edge of the blade. Next, cut off the two rough faces from each leg. Finish machining of each leg by thickness planing or jointing it to 3" square.

3 Crosscut the legs, squaring both ends for a finished length of 30¾".

Note: Legs of this length make a workbench that stands 33" high. Lengthen or shorten the legs to suit your personal preference.

4 Put a ½" dado blade in your table saw and adjust for a ½"-deep cut. Set your rip fence 1¼" from the inside edge of the blade, and cut grooves centered in two adjacent faces of each leg, where shown on Drawing-9.

5 Mark the centerpoints of all holes in the legs, where shown on the Drawing 8.

Note: The faces of the two legs on the side with the side vise are the only ones that get the row of support-pin holes spaced 3" on center. Chuck a 1" Forstner bit into your drill press, then set up a fence so the bit is centered in the width of the leg. Drill all the 1" holes and counterbores where dimensioned. Then, change to a 13/32" bit, and complete the holes through the legs.

6 Chuck a chamfering bit into your hand-held router, and adjust it to cut a ⅛" chamfer. Rout the edge of all 1" holes in the legs. Next, chuck the chamfering bit into your table-mounted router, and adjust it to cut a ¼" chamfer. Then, rout the edges and bottom ends of all the legs.

Note: Save this router-table setup; you will use it again. Sand the legs to final smoothness, and set them aside.

Next, shape the aprons and-stretchers

1 Rip and crosscut the aprons, (H,-I) and the stretchers (J, K) from 1½"-thick stock (we used construction-grade fir) to the dimensions listed in the Materials List.

2 Put a ⅝" dado blade in your table saw, and raise it ½" above the table. Attach an extension to your table saw's miter gauge, and clamp a stopblock to it to make a cut ½" wide. (Refer to Drawing 9a.) Make test cuts in a piece of scrap stock, and then test the fit of the tenon in the groove in the leg. Next, cut tenons on the ends of the aprons and stretchers (H, I, J, K). Change to a ½" dado blade in your table saw, and cut grooves along the bottom edge of each apron and stretcher.

3 Mark the centerpoints of the 1" shelf-rod holes in the side stretchers (J).

Note: The first hole is 1½" from the end of the side stretcher (1" from the shoulder of the tenon).

Chuck a 1" Forstner bit into your drill press, and then clamp a fence 2" from the center of the bit. Next, drill a hole in scrap wood, and test the fit of your 1" dowel stock. You want a loose fit; these dowels will not be glued into place. Then, drill the holes.

4 Crosscut 12 pieces of 1" dowel stock to 17¼" long for the shelf rods. Crosscut two 3½" lengths for the support pins. Sand the dowels, and then set aside.

5 Rout ¼" chamfers along the bottom edge (with the groove) of the end aprons (I). Next, rout ¼" chamfers along all long edges of parts H, J, and K.

> **SHOP TIP**
>
> To avoid ruining the thread on all-thread rod when hacksawing to length, screw a nut onto the rod, and then cut the rod. Remove the nut to re-form the threads.

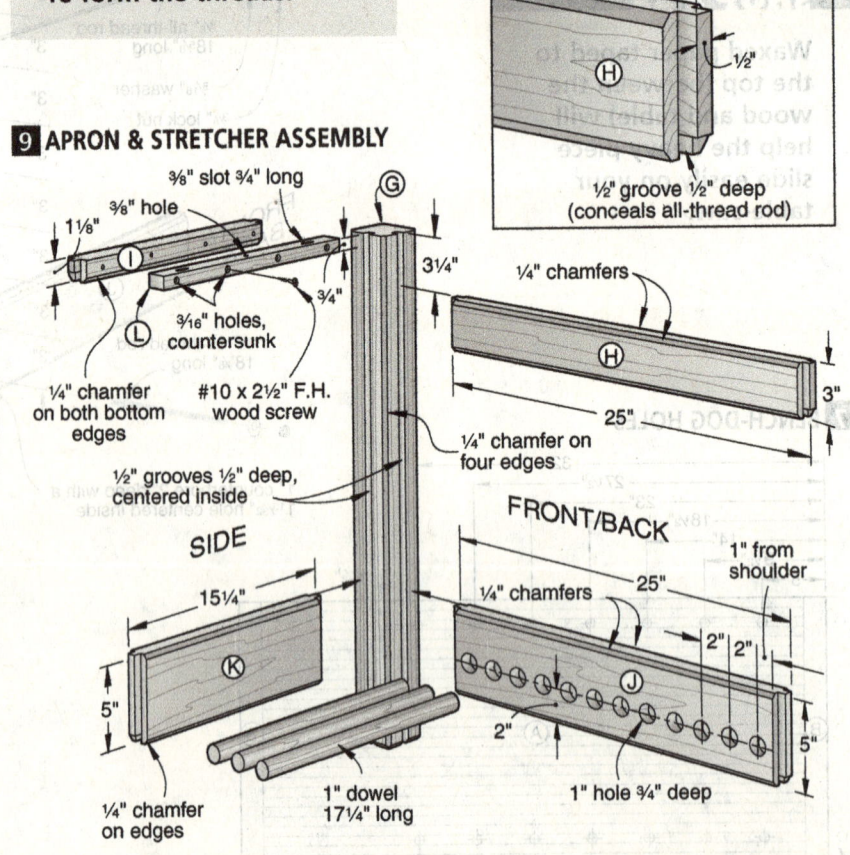

9a TENON & GROOVE DETAIL

9 APRON & STRETCHER ASSEMBLY

214 Project 6

Article adapted from WOOD magazine, Weekend Woodworking Projects, September 1995

10 PARTS VIEW

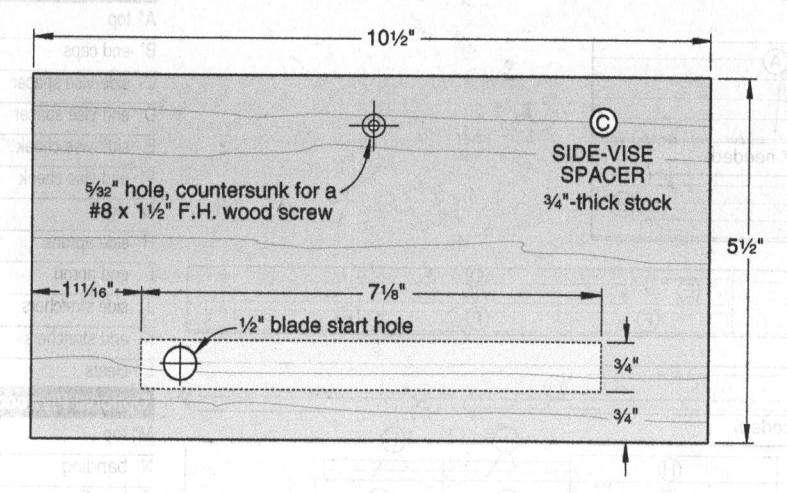

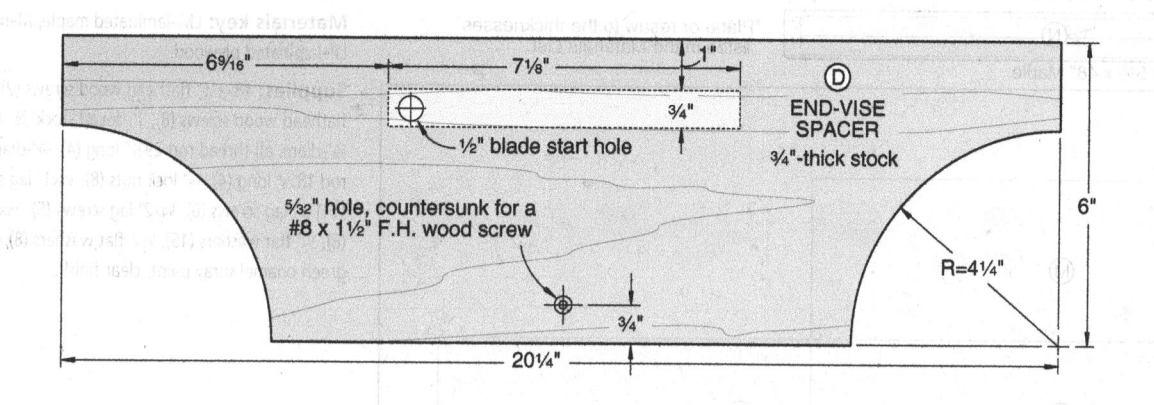

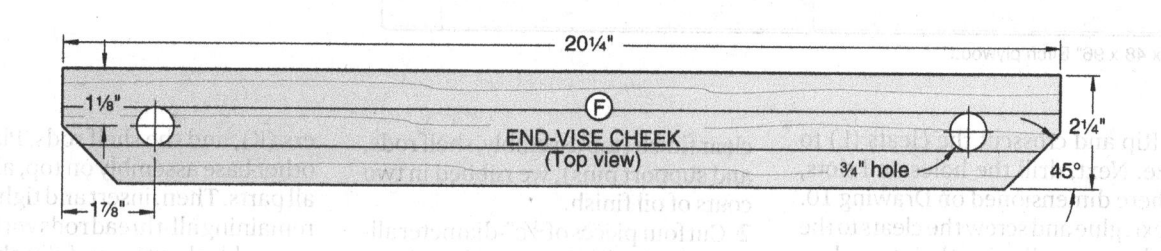

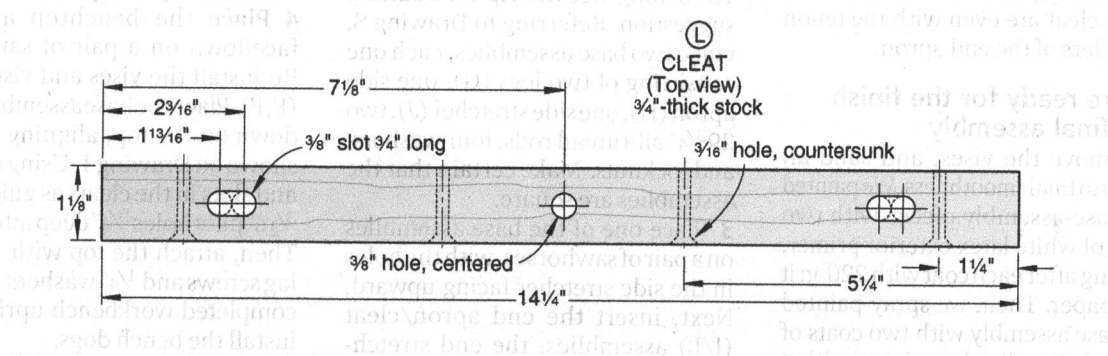

C — SIDE-VISE SPACER
¾"-thick stock
- 10½" × 5½"
- 5/32" hole, countersunk for a #8 × 1½" F.H. wood screw
- ½" blade start hole
- 1 11/16", 7⅛", ¾", ¾"

D — END-VISE SPACER
¾"-thick stock
- 6 9/16", 7⅛", 1"
- ½" blade start hole
- 5/32" hole, countersunk for a #8 × 1½" F.H. wood screw
- ¾"
- 20¼" × 6"
- R = 4¼"

F — END-VISE CHEEK (Top view)
- 20¼"
- 1⅛", 1⅞"
- ¾" hole
- 2¼"
- 45°

L — CLEAT (Top view)
¾"-thick stock
- 7⅛"
- 2 3/16", 1 13/16"
- ⅜" slot ¾" long
- 3/16" hole, countersunk
- ⅜" hole, centered
- 1⅛"
- 14¼"
- 5¼", 1¼"

Article adapted from WOOD magazine, Weekend Woodworking Projects, September 1995

Project 6 215

Cutting Diagram

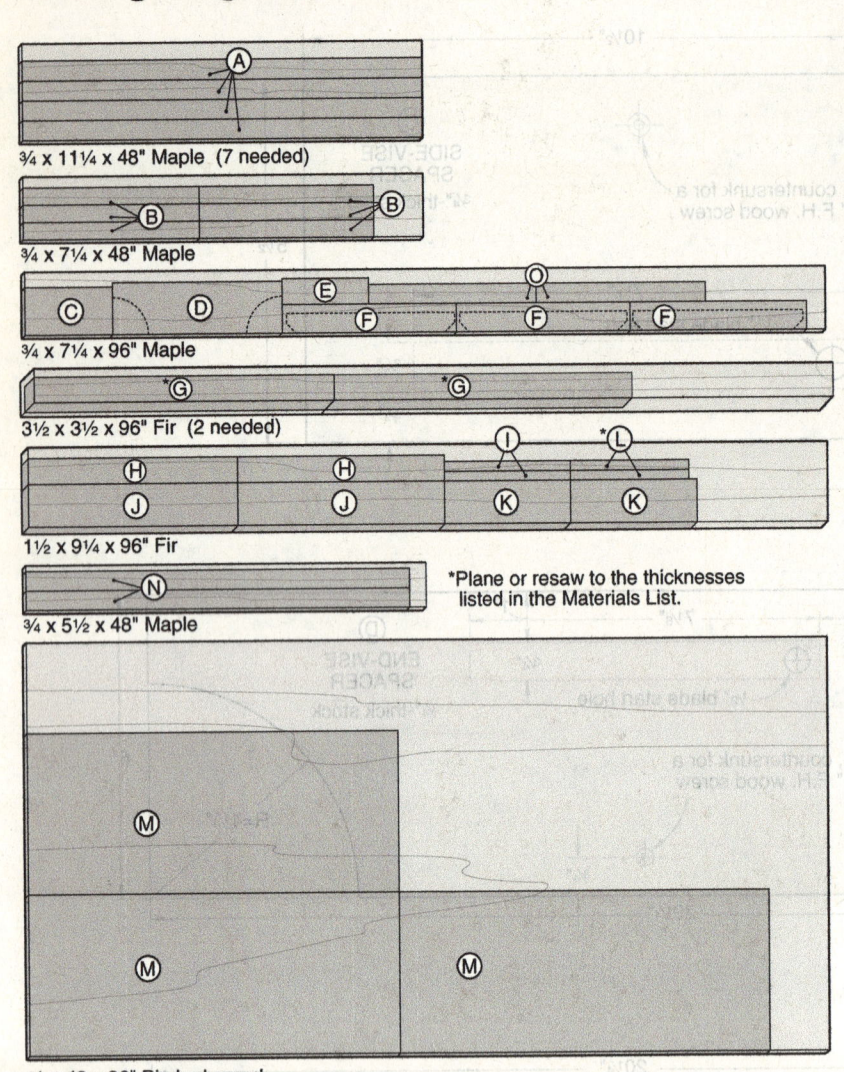

¾ x 11¼ x 48" Maple (7 needed)

¾ x 7¼ x 48" Maple

¾ x 7¼ x 96" Maple

3½ x 3½ x 96" Fir (2 needed)

1½ x 9¼ x 96" Fir

¾ x 5½ x 48" Maple

*Plane or resaw to the thicknesses listed in the Materials List.

¾ x 48 x 96" Birch plywood

Materials List

Workbench	T	W	L	Matl.	Qty
A* top	2¼"	20¼"	44"	LM	2
B* end caps	2¼"	1¾"	20¼"	LM	2
C side-vise spacer	¾"	5½"	10½"	M	1
D end-vise spacer	¾"	6"	20¼"	M	1
E side-vise cheek	¾"	3"	10½"	M	1
F* end-vise cheek	2¼"	3"	20¼"	LM	1
G* legs	3"	3"	30¾"	F	4
H side aprons	1½"	3"	25"	F	2
I end apron	1½"	1⅛"	15¼"	F	2
J side stretchers	1½"	5"	25"	F	2
K end stretchers	1½"	5"	15¼"	F	2
L cleats	¾"	1⅛"	14¼"	F	2
Optional Plywood Top					
M* top	2¼"	18¾"	44½"	LP	1
N banding	¾"	2¼"	46"	M	2
O banding	¾"	2¼"	20¼"	M	2

*Parts initially cut oversize. See the instructions.

Materials key: LM–laminated maple, M–maple, F–fir, LP–laminated plywood.

Supplies: #8x1½" flathead wood screws (2), #10×2½" flathead wood screws (8), 1" dowel stock 36" long (7), ⅜"-diam. all-thread rod 29¾" long (4), ⅜"-diam. all-thread rod 18⅝" long (4), ⅜" lock nuts (8), ¼x1" lag screws (4), ¼x1½" lag screws (6), ¼x2" lag screws (5), ⅜x2" lag screws (8), ¼" flat washers (15), 5/16" flat washers (8), primer, green enamel spray paint, clear finish.

---

6 Rip and crosscut the cleats (L) to size. Next, drill the holes and slots, where dimensioned on Drawing 10. Next, glue and screw the cleats to the end aprons, aligning their top edges and making certain that the ends of the cleat are even with the tenon shoulders of the end apron.

You're ready for the finish and final assembly

1 Remove the vises, and sand all pieces to final smoothness. We painted the base-assembly pieces with two coats of white latex exterior primer, sanding after each coat with 220-grit sandpaper. Then, we spray-painted the base assembly with two coats of enamel. On all other pieces with a clear finish (top assembly, shelf rods, and support pins), we rubbed in two coats of oil finish.

2 Cut four pieces of ⅜"-diameter all-thread rod 29¾" long, and four pieces 18⅝" long. See the Tip for a cutting suggestion. Referring to Drawing 8, make two base assemblies, each one consisting of two legs (G), one side apron (H), one side stretcher (J), two 29¾" all-thread rods, four washers, and locknuts. Make certain that the assemblies are square.

3 Place one of the base assemblies on a pair of sawhorses, with the holes in the side stretcher facing upward. Next, insert the end apron/cleat (I/L) assemblies, the end stretchers (K), and the shelf rods. Place the other base assembly on top, aligning all parts. Then, insert and tighten the remaining all-thread rods with washers and locknuts, carefully checking the assembly for square.

4 Place the benchtop assembly facedown on a pair of sawhorses. Re-install the vises and vise cheeks (E,F). Place the base assembly upside down on the top, aligning it where shown on Drawing 1. Using the holes and slots in the cleats as guides, drill 3/16" pilot holes ¾" deep into the top. Then, attach the top with ¼x1½" lag screws and ¼" washers. Turn the completed workbench upright, and install the bench dogs.

Big Rig for Little Drivers

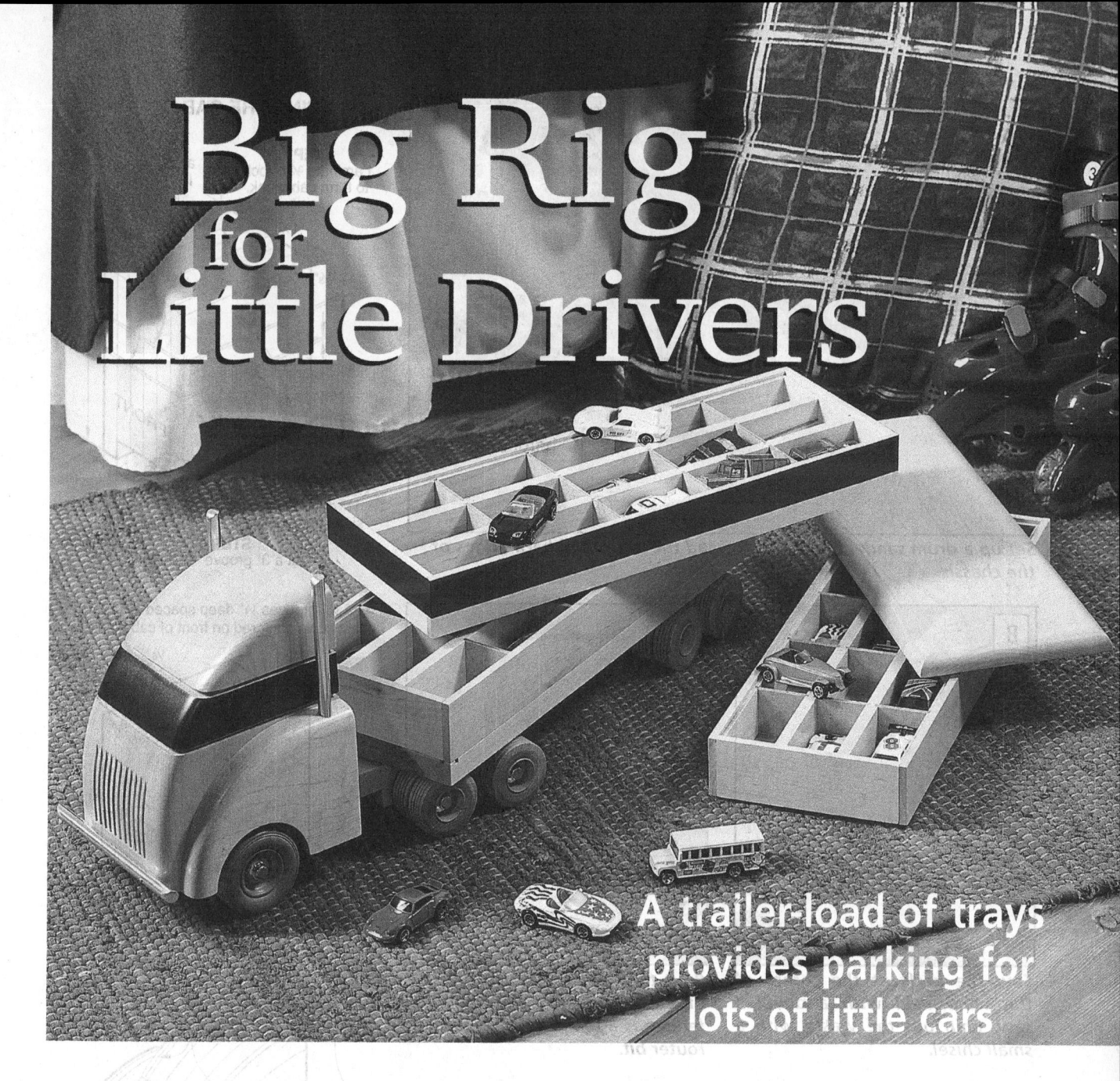

A trailer-load of trays provides parking for lots of little cars

Little cars and trucks pose a big problem: Finding a place to park them after playtime. This truck delivers a solution: Its triple-tray trailer holds 45 of the popular 3" vehicles.

Cut out the truck's chassis first

1 Cut the cab chassis (A) to size. Bandsaw the narrowed rear section of the chassis, where shown on the Cab Chassis Top View drawing. Saw slightly outside the line; then sand to the line with a ⅝" drum sander and a fence as shown in *Photo A*. Set the fence ⅛" behind the edge of the drum, and attach a strip of ⅛" hardboard to the outfeed side of the fence, using double-faced tape.
2 Drill and chamfer the trailer kingpin hole on top of the chassis.

Form the front fenders next

1 Cut two ¾ × 6 × 5½" blanks for the fenders (B), and temporarily laminate them with double-faced tape. Adhere the Fender Full-Size Pattern (in the pattern section at the end of this article) to the stack, placing the top of the fender outline flush with the top edge.
2 Bore the 3" hole and rip the workpiece to 4" wide where shown.
3 Drill a ⅜" exhaust-stack hole centered on the top edge of each fender where shown.
4 Bandsaw the outline, and sand to the pattern line. Rout a ½" roundover along the outer edge of each fender. Separate the fenders.

Now, construct the cab

1 Laminate six ¾ × 7⅜ × 6" pieces of stock for the cab blank, as shown in *Step 1* of the Forming the Cab drawing. Attach the Full-Size Cab Pattern to one face.
2 Saw kerfs across the front, sides, and back, where shown in *Steps 2 and 3* of the drawing.
3 When cutting the groove in *Step 4* of the drawing, clamp a scrapwood backing piece to the blank to prevent tearout on the exit side of the cut.

Article adapted from WOOD magazine, issue 118, November 1999

Project 7

A

Set up a drum sander and fence to sand the narrowed back edges of the chassis.

B

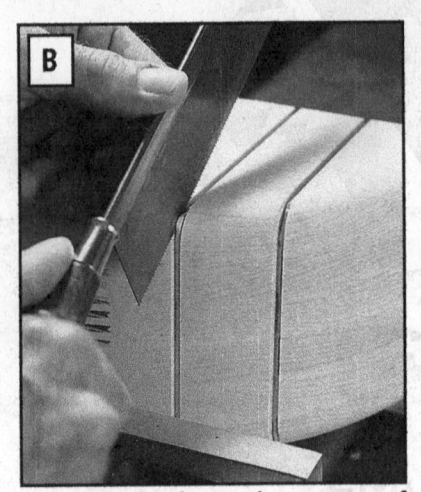

Saw the kerfs at the corner of the windshield with a dovetail saw. Chip out the waste with a small chisel.

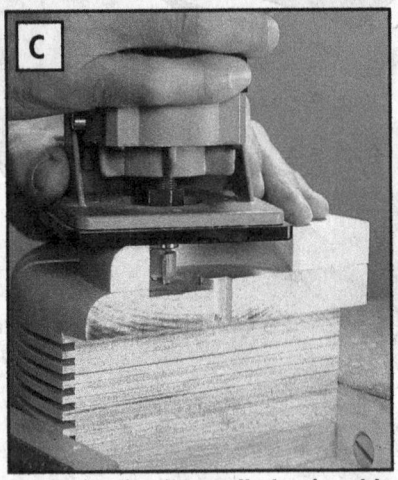

C

Rout the fender wells in the side of the cab using the installed fender as a template for a pattern router bit.

FORMING THE CAB

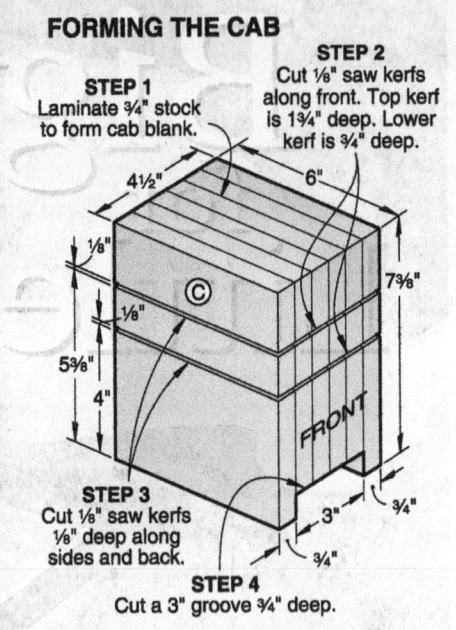

STEP 1 Laminate ¾" stock to form cab blank.

STEP 2 Cut ⅛" saw kerfs along front. Top kerf is 1¾" deep. Lower kerf is ¾" deep.

STEP 3 Cut ⅛" saw kerfs ⅛" deep along sides and back.

STEP 4 Cut a 3" groove ¾" deep.

STEP 5 Cut ⅛" grooves ¼" deep spaced ⅛" apart. First cut is centered on front of cab blank.

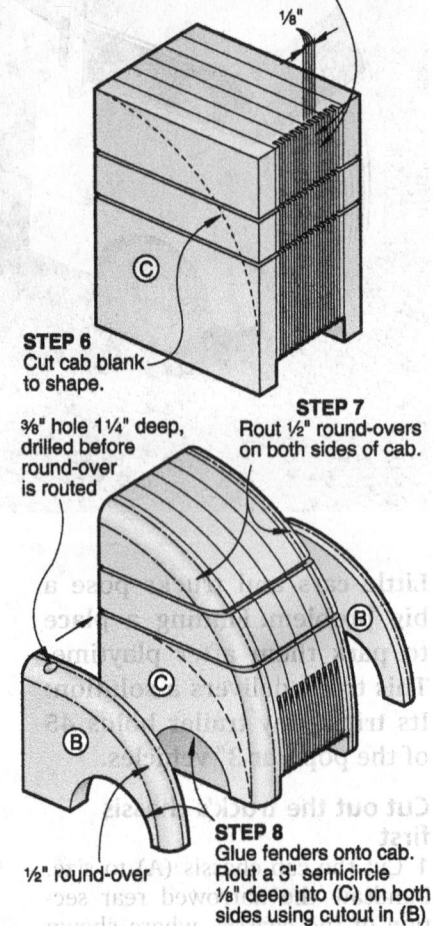

STEP 6 Cut cab blank to shape.

STEP 7 Rout ½" round-overs on both sides of cab.

STEP 8 Glue fenders onto cab. Rout a 3" semicircle ⅛" deep into (C) on both sides using cutout in (B) as a guide.

⅜" hole 1¼" deep, drilled before round-over is routed

½" round-over

4 When you get to *Step 7* of the drawing, rout the round-overs on the cab sides with a table-mounted router. Set a fence flush with the round-over bit's pilot to prevent the bit from falling into the window kerfs when routing.

After routing the round-overs, cut back the corners of the windshield kerfs, as shown in *Photo B*.

5 After gluing the fenders to the cab sides in *Step 8*, rout the wheel well on each side ⅛" deep into the side of the cab with a handheld router and a pattern bit (*Photo C*).

6 Glue the chassis (A) into the groove in the cab (C), bringing the parts flush at the front. Rout a ¼" round-over along the bottom front.

7 Cut a 13/32" dado 7/32" deep across the bottom of the cab/chassis assembly, where shown in the Cab Chassis drawing. Verify that the dimension shown centers the dado in the assembly's fender openings.

8 Cut the front chassis block (D) and rear chassis block (E) to the dimensions shown in the Bill of Materials. Referring to the Front Chassis Block drawing, form a dado to mate with the one in the chassis. Drill two 13/32" holes in part E, where shown in the Rear Chassis Block drawing.

9 Glue and clamp parts D and E in place. Insert a piece of ⅜" rod to maintain alignment between the dadoes in part A and D.

218 Project 7

Article adapted from WOOD magazine, issue 118, November 1999

Big Rig

Now, make a flatbed trailer
1 Cut the trailer bed (F) to size.
2 Drill a 3/8" hole 1/2" deep in the bottom of the trailer bed, where shown on the Trailer Bed drawing.
3 Saw the 1/8" kerfs on top.
4 Cut parts G and H to the dimensions shown. Fasten the trailer blocks together face-to-face with double-faced tape, and drill the 13/32" holes, where shown in the Trailer Chassis drawing. Separate the parts.
5 Glue and clamp the trailer chassis (G) and trailer blocks (H) together. Keep the ends and top surfaces flush.
6 Glue and clamp the assembly to the bottom of the bed where shown.

Now, add the cargo containers
1 Cut the tray sides (I) and ends (J) to the dimensions shown.
2 Saw or rout a 1/4" rabbet 1/8" deep along the bottom outside edge of each side (I) and end (J), shown on the Tray Exploded View drawing. The resulting tongues should fit loosely into the grooves in the trailer bed (F).
3 Cut a 1/8" rabbet 1/8" deep along the inside top edge of each side (I) and end (J).
4 Cut 1/8" grooves for the bottoms in the sides (I) and ends (J).
5 Cut 3/8" rabbets 1/8" deep in the ends of the tray sides (I) as shown.
6 Cut 3/8" dadoes 1/8" deep in the tray ends (J) where shown.
7 Cut the tray bottoms (K) to size. Dry-clamp the tray assemblies (I, J, K). Test each one to ensure it fits into the grooves in the trailer bed (F). Check that the trays stack together easily.
8 Measure the trays to determine the length for the tray dividers (L). Cut the parts to size.
9 Unclamp the trays, and dado the sides (I) and tray dividers (L) for the partitions (M). Attach an auxiliary fence to your table saw's miter gauge, and clamp a stopblock to it to make each cut on all parts.
10 Glue together and clamp the tray assemblies (I, J, K, L). To ensure that each one is flat and square, assemble the trays one at a time, and clamp each into the grooves in the trailer

Bill of Materials

Part	T	W	L	Matl.	Qty.
A cab chassis	3/4"	3"	12 5/8"	M	1
B* fender	3/4"	4"	5 1/2"	M	2
C cab	4 1/2"	6"	7 3/8"	LM	1
D front block	5/8"	4 1/4"	2"	M	1
E rear block	5/8"	2 3/4"	4 3/4"	M	1
F trailer bed	3/4"	6"	16 7/8"	M	1
G trailer chassis	3/4"	1 1/4"	4 3/4"	M	1
H trailer block	3/4"	1 5/8"	4 3/4"	M	2
I tray side	3/8"	2"	16 7/8"	M	6
J tray end	3/8"	2"	5 1/2"	M	6
K tray bottom	1/8"	5 1/2"	16 3/8"	H	3
L** tray divider	3/8"	1 1/2"	16 3/8"	M	6
M tray partition	1/8"	1 1/2"	1 3/4"	H	36
N top	3/4"	6"	16 7/8"	M	1

* Start with oversized blanks to make fenders; see instructions.
** Cut part longer initially; then trim to finished length in accordance with instructions.

Materials Key: H–tempered hardboard; M–soft maple; LM–laminated soft maple

Supplies: 3/8" aluminum rod, 3/8" push-on axle caps, #16 × 1" brass escutcheon pins, 2" dual wheels, 2 1/2" wheels, epoxy glue.

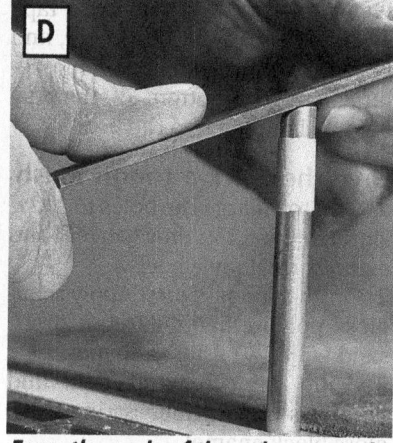

Form the ends of the exhaust stacks and bumper with a file.

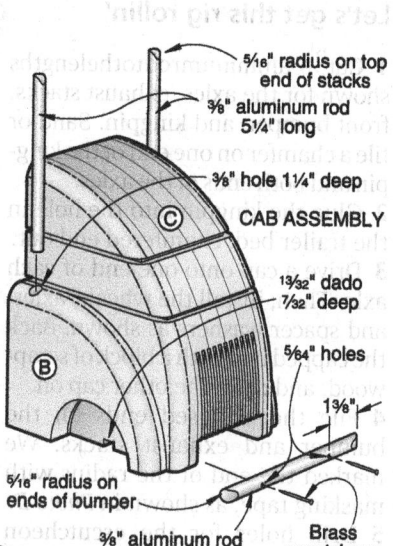

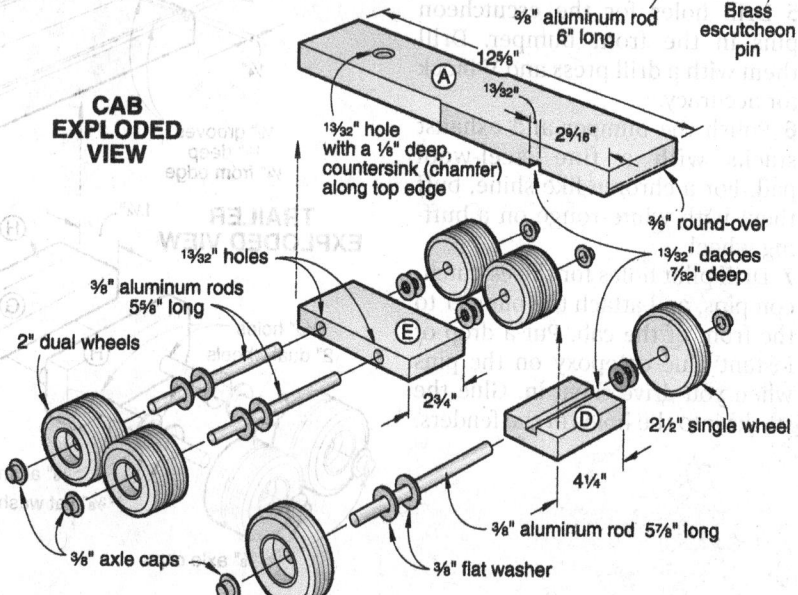

CAB ASSEMBLY

CAB EXPLODED VIEW

Article adapted from WOOD magazine, issue 118, November 1999

Project 7 219

bed (F). Put strips of masking tape along the sides of the grooves, and be sure to clean off all glue squeeze-out along the bottom of the tray before clamping it to the bed.

Give the truck a tough finish

1 Drill to enlarge the holes to $^{13}/_{32}$" through two $2^{1}/_{2}$" toy wheels and eight 2" dual toy wheels.
2 Finish-sand all parts. Apply semi-gloss polyurethane. After it dries, sand with 320-grit sandpaper.
3 Paint the cab windshield and windows and trailer stripe with gloss black enamel.
4 After the enamel dries, mask over it. Then, apply a second coat of clear finish to all parts. Allow to dry.
5 Cut the partitions (M) to size. Fit them into the dadoes in the trays, but do not glue them in place.

Let's get this rig rollin'

1 Cut $^{3}/_{8}$" aluminum rod to the lengths shown for the axles, exhaust stacks, front bumper, and kingpin. Sand or file a chamfer on one end of the kingpin and both ends of the axles.
2 Glue the kingpin into the hole in the trailer bed, chamfered end out.
3 Drive a cap onto one end of each axle. Then, install the wheels, axles, and spacer washers as shown. Back the capped end with a block of scrapwood, and drive the other cap on.
4 File the radiused ends on the bumper and exhaust stacks. We marked the end of the radius with masking tape, as shown in *Photo D*.
5 Drill holes for the escutcheon pins in the front bumper. Drill them with a drill press and V-block for accuracy.
6 Polish the bumper and exhaust stacks with a fine steel-wool pad. For a chrome-like shine, buff them with white rouge on a buffing wheel.
7 Drill pilot holes for the escutcheon pins, and attach the bumper to the front of the cab. Put a drop of instant glue or epoxy on the pins when you drive them in. Glue the stacks into the holes in the fenders.

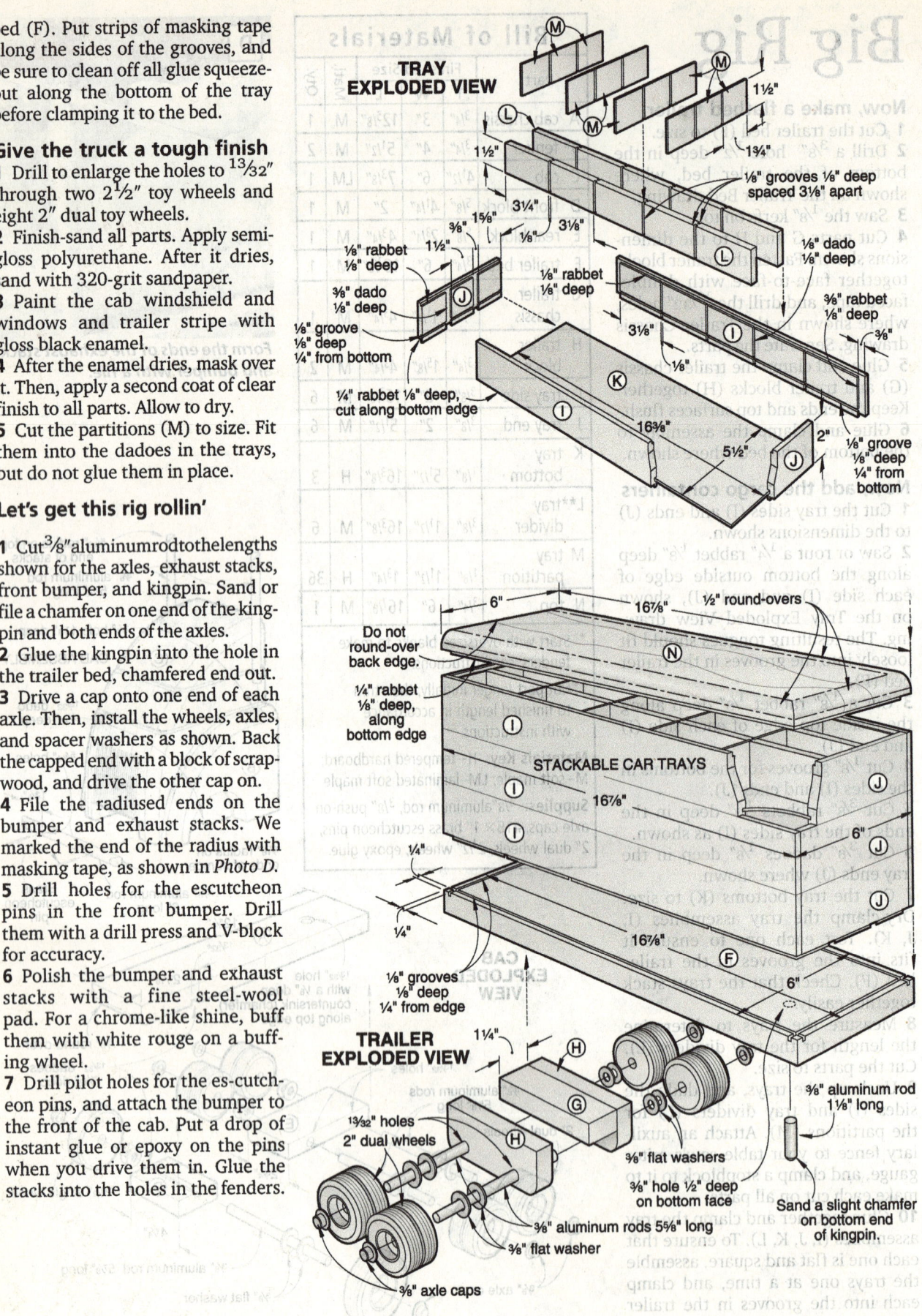

Patterns for Big Rig

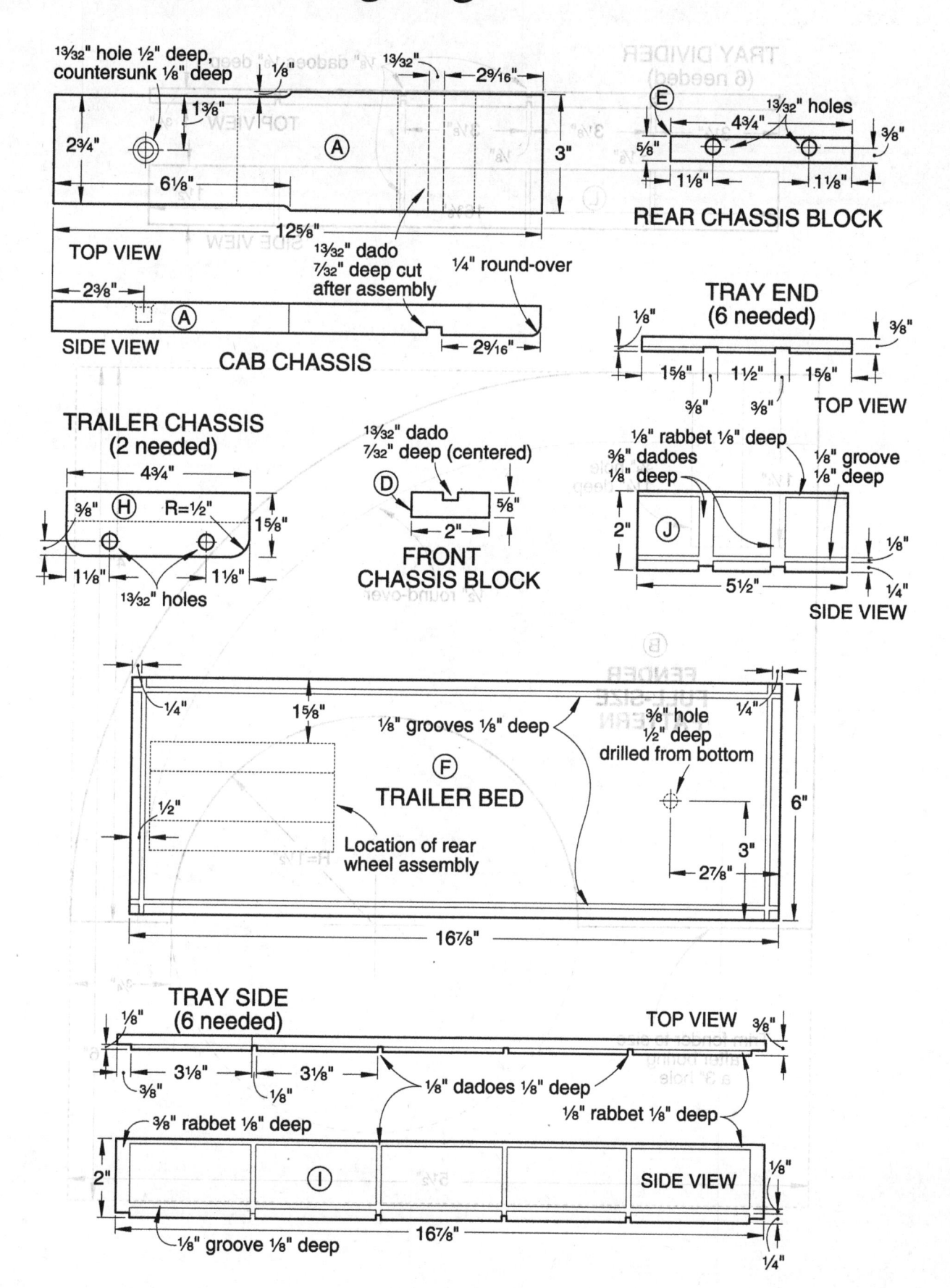

Patterns for Big Rig

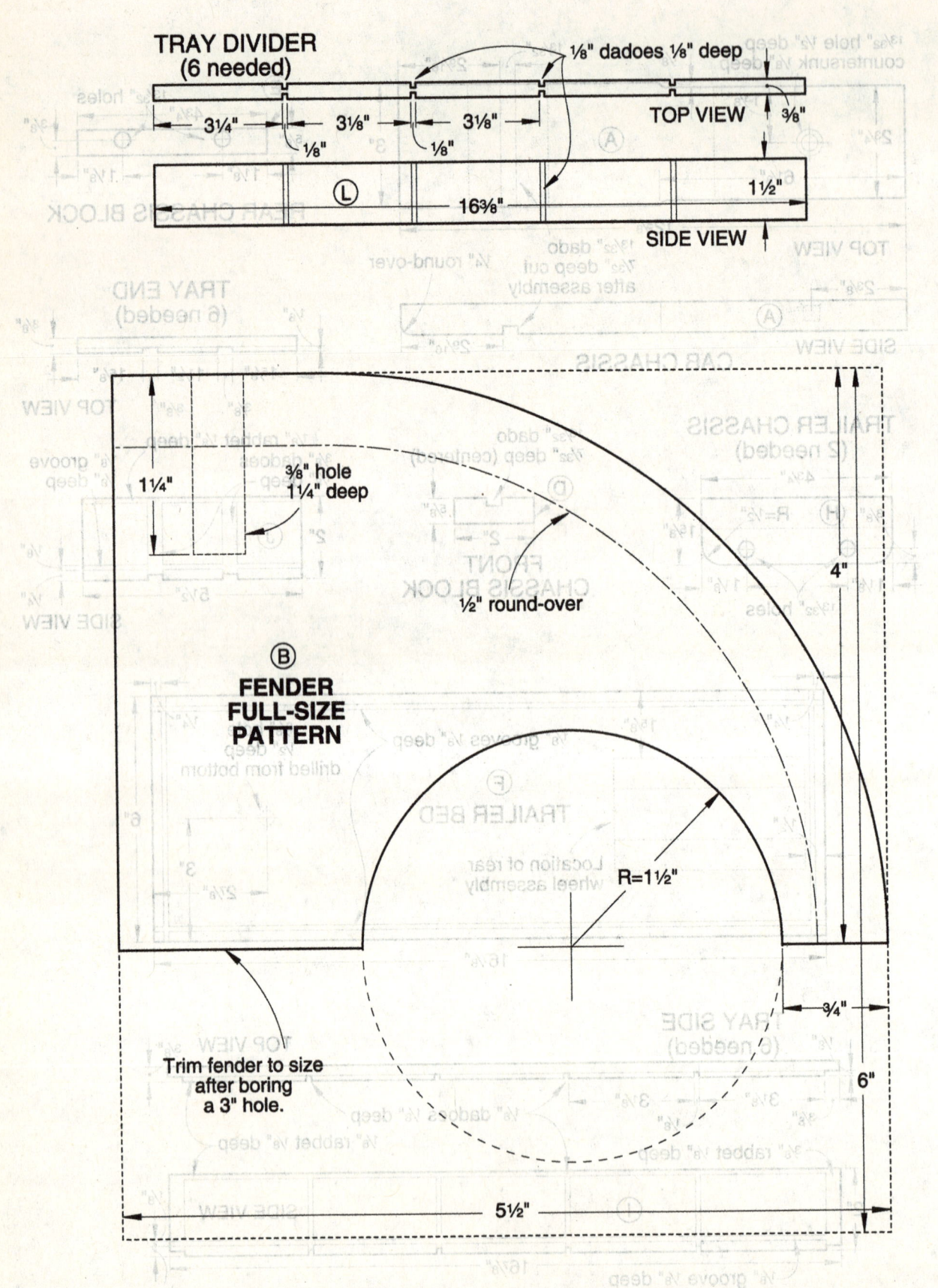

TRAY DIVIDER
(6 needed)

⅛" dadoes ⅛" deep

3¼" — 3⅛" — 3⅛"

⅛" — ⅛"

TOP VIEW ⅜"

Ⓛ 16⅜" 1½"

SIDE VIEW

1¼" ⅜" hole 1¼" deep

½" round-over

Ⓑ **FENDER FULL-SIZE PATTERN**

R=1½"

4"

¾"

6"

Trim fender to size after boring a 3" hole.

5½"

222 Project 7

Article adapted from WOOD magazine, issue 118, November 1999